T0275615

**Waste Is Information**

**Infrastructures Series**

edited by Geoffrey C. Bowker and Paul N. Edwards

# Waste Is Information

## Infrastructure Legibility and Governance

Dietmar Offenhuber

foreword by Carlo Ratti

The MIT Press
Cambridge, Massachusetts
London, England

© 2017 Massachusetts Institute of Technology

All rights reserved. No part of this book may be reproduced in any form by any electronic or mechanical means (including photocopying, recording, or information storage and retrieval) without permission in writing from the publisher.

This book was set in ITC Stone Sans Std and ITC Stone Serif Std by Toppan Best-set Premedia Limited.

Library of Congress Cataloging-in-Publication Data

Names: Offenhuber, Dietmar, author.
Title: Waste is information : infrastructure legibility and governance / Dietmar Offenhuber ; foreword by Carlo Ratti.
Description: Cambridge, MA : MIT Press, [2017] | Series: Infrastructures |
    Includes bibliographical references and index.
Identifiers: LCCN 2017000879 | ISBN 9780262036733 (hardcover : alk. paper)
ISBN 9780262549967 (paperback)
Subjects: LCSH: Refuse and refuse disposal--Planning. | Refuse and refuse
    disposal--Citizen participation. | Municipal services--Planning--Case
    studies. | Refuse and refuse disposal--Philosophy. | City planning.
Classification: LCC TD793 .O45 2017 | DDC 363.72/8--dc23 LC record available at
https://lccn.loc.gov/2017000879

# Contents

# Foreword

**Carlo Ratti**

What will the human race be at the moment of its extinction? A certain quantity of information about itself and the world, a finite quantity, given that it will no longer be able to propagate itself and grow.

—Italo Calvino, *World Memory and Other Cosmicomic Stories*, 1968

A little-known short story by Italian writer Italo Calvino, called "World Memory," imagines a future human condition when all information produced by humanity will be stored and available for search. In other terms, it will be "an archive that will bring together and catalogue everything that is known about every person, animal and thing, by way of a general inventory not only of the present but of the past too, of everything that has ever been since time began, in short a general and simultaneous history of everything, or rather a catalogue of everything, moment by moment."

Calvino's fictional condition could not be more topical today, as the amount of digital data we generate and store on the planet is growing at an unprecedented rate. It is sometimes estimated that up to 90 percent of all of the information available in the world today was produced in the past two years (2015 and 2016) alone. Most of our actions—the calls we make, the apps we use, the credit cards we scan—are recorded and added to a growing digital repository of human life. Our deliberate actions also contribute toward this ever-growing Big Data: when we post a picture, tweet a thought, or share moments of our life with friends and the broader network. To employ a word now commonly adopted in research, our own human self is becoming quantifiable and quantified.

Most of the digital footprints that are recorded today are somehow connected with some measure of human activity. They tend to concentrate in cities (where humans abound …) and record actions that society considers of value—things that are meaningful to us. In this book, Dietmar Offenhuber takes an unprecedented approach and turns his informational lens upside down. He focuses not on what is meaningful and has value but on what society discards: waste. He uses it as the starting point of a reflection on human life not through its direct observation, but through the proxy of what it constantly excretes.

As archaeologists have often shown, dumps can offer unprecedented snapshots of past civilization. The minutia of life that might be lost in official recordings emerges there with new force. The immediacy of what has been disposed of provides unexpected angles on the processes that produced it—and then got rid of it as waste. Also, the examination of what is discarded, and what is not, helps us also understand what was "waste" and what was not, a definition that changes in different times and different societies. Offenhuber takes this lens to the present day and—as a digital archaeologist of the contemporary era—embarks in a multifaceted analysis of "Waste as Information."

Implicit in Offenhuber's investigation seems to be an agenda of social justice. Waste is just another aspect of societal rejection, as determined by the local conventions of a given time and place. He writes: "Waste systems cannot be separated from systems of production, and notions of value cannot be seen in isolation." The value of an object and the value of the human work attached to it cannot be separated. By elevating waste as the focus of this investigation, Offenhuber seems to rehabilitate those circles of society that suffer from the same rejection.

As one reads through the pages of this fascinating book, little by little the line between waste and nonwaste becomes increasingly blurred. This reminds us of an excerpt from Italo Calvino's *Invisible Cities* (1972), where waste seems to become the organizing principle of the world in the city of Leonia: "Leonia's rubbish little by little would invade the world, if, from beyond the final crest of its boundless rubbish heap, the street cleaners of other cities were not pressing, also pushing mountains of refuse in front of themselves. Perhaps the whole world, beyond Leonia's boundaries, is covered by craters of rubbish, each surrounding a metropolis in constant eruption."

# Preface: The Paper Police

All that is necessary already exists ... it is just wrongly distributed.
—Jürgen O. Olbrich, e-mail interview with author, March 3, 2010

For the past twenty years, Kassel-based artist Jürgen O. Olbrich has been scavenging paper and cardboard containers looking for printed material—books, manuals, documents, postcards, and reports. Influenced by the Fluxus movement's rejection of the separation between art and everyday life, his excursions are informed by the goal of losing nothing worth preserving. He is motivated by how everyday culture and history gets lost because the importance of discarded items has been forgotten.

Strictly speaking, Olbrich's activity might be considered theft. Depending on local ordinances, materials in containers are often regarded as property of the communal recycling company. However, Olbrich has never encountered legal trouble. During his collection trips, he wears a uniform bearing the name Paper Police. By associating the theme of enforcement with his illicit activity, he has, comically and ironically, avoided confrontation. People show him respect during chance encounters at recycling receptacles. Many ask if he has lost something, though he might ask them the same thing.

Olbrich is fascinated by the number of personal documents people discard, oblivious to the possibility that someone else might find them. In the areas around senior housing, he has found documents and keepsakes that chronicle entire lives. In public and busy locations, he frequently finds material that is potentially dangerous or embarrassing for the previous owner, such as bank account information or personal pornography.

Since starting the project in 1989, Olbrich has diligently refined his methods, hiring employees who receive detailed instructions about his foraging methods, which entail distinguishing between rare and abundant items and making quick decisions about what to keep or discard. He avoids categorizing the items he keeps, but sometimes collects according to themes like paraphernalia about Kassel, perfume-bottle boxes, or contact sheets of photo negatives. Items Olbrich has found reside in fifteen private and public collections, some of which are dedicated to specialized topics, such as a research project on 1930s German tourist brochures.

Olbrich does not see himself as a collector; he does not strive for completeness. He is driven by the anticipation of a lucky find. Sometimes he does find valuable or otherwise noteworthy documents like a letter from Albert Einstein, a 1969 art book from the American artist Ed Ruscha, or a collection of birthday greeting cards from members of the German Bundestag. He often wraps the objects he finds in official "Paper Police" gift wrap, giving the packages to people he encounters to remind them "of what keeps us occupied and what we love." When I met him in 2008 in Linz, Austria, he gave me a package that bundled a declassified intelligence report about the German Democratic Republic with a collection of recipes involving Jägermeister liquor.

Through this evolving project, Olbrich has learned a thing or two about his territory. He knows the locations of the most bountiful recycling bins and their weekly collection schedules. He has become acquainted with other foragers who specialize in antiquarian books. He has learned what to look for in certain areas, and associates each neighborhood with characteristic items of waste that he turns into something else: evidence, cultural artifacts, works of art.

There are many like Olbrich who read waste systems in their own way—foragers, gatherers, and trackers of material traces or digital data. Waste pickers scavenging reusable materials, investigative reporters searching for evidence of illegal waste exports, manufacturers trying to understand their supply chains, sanitation departments measuring system performance, or neighbors fighting a proposed transfer station all read waste systems differently, and they all use their specific methods of representation.

This book is an invitation to consider waste systems from the perspective of information. That is, it looks at how we read, experience, and investigate urban services, which data collection methods and visualizations we use, and what roles these representations play in the governance of infrastructural systems.

# Introduction: Waste Is Information

At the present time the conditions of town life have changed. We can no longer utilize most of the rejected food matter for feeding, except in small towns. We now discard more rubbish, our fuel varies more as to its kind and the quantity used, we can no longer use the older and crude methods of collection and delivery, we have materially increased the distances to which refuse must be delivered and we are abandoning the disposal by dumping except with ashes. The problem of refuse disposal has, therefore, become more complex than formerly, and this complexity may not yet have reached its limit.

In order that correct solutions for the best methods of disposal may be found, both from the standpoint of sanitation and economy, it is necessary to inquire into details far more than formerly, so as to have more definite facts and figures with which to solve the problem. ... In short, we must have more special data and statistics before we can indicate the best methods for the disposal of a particular town's refuse.

At the outset we should know what the refuse consists of, and ascertain and discriminate between its various parts, which may be enumerated as being garbage, dead animals, night soil, manure, street sweepings, ashes and rubbish.

—Rudolf Hering (1912, 909)

Waste has been described as a nuisance, a threat, a source of injustice, and a symptom of excess. Considering the overwhelming materiality of garbage and the urgency of environmental concerns, describing it as "information" might seem inappropriate. But studying waste from an informational perspective does not imply that it is immaterial, or that problems of pollution and overconsumption are imaginary. In some sense, waste is physically embodied information. Much of what we know about the past, we know from things thrown away, the discarded objects that record human activity and the passage of time. Historical dumpsites are of interest to archaeologists and anthropologists, just as contemporary landfills harbor a wealth of information about everyday consumption and behavior (Rathje and Murphy 2001).

Waste can be information in the literal sense. Until a few years ago, data generated by the data infrastructures of telephone providers, online platforms, or traffic control systems were seen as transient, unwieldy, and not worth storing—informational waste. Under the banner of Big Data, these leftovers are now praised as "the new oil" in *Forbes* magazine. This "digital exhaust" is mined by machine-learning algorithms that try to recognize patterns, classify items, and predict behaviors. The data sets are often collected for unrelated purposes, arrive in incompatible formats, and are stored without a strict data structure—making the data centers that host them a bit like landfills themselves. After all, actual landfills are increasingly mined for methane, metals, and other valuable commodities (Jones et al. 2013).

Viewing waste as information stems from the realization that waste is, above all, a designation. Waste is whatever is labeled as waste, and nothing exists that cannot become waste at some point. In his last book, the posthumously published *Wasting Away*, urban designer Kevin Lynch defines "waste" as a lost opportunity, "a resource not in use, but potentially useful: wasted time, a wasted life, an empty building or field, an idle machine" (Lynch 1991). The European Union defines "waste" as "any substance or object which the holder discards or intends or is required to discard" (European Commission 2008).

"Waste" is defined through context, relative to a value system. The anthropologist Mary Douglas observed, "Where there is dirt there is system." As she argues, dirt is a byproduct of ordering and classification, and dirt is what violates this symbolic order (Douglas 1966, 41). In information systems, "dirt" often refers to erroneous data entries or classification errors. Studying such inconsistencies can reveal how data were produced (Loukissas 2017).

In the more prosaic language of economics, Richard Porter defines "waste" plainly as "anything that is no longer privately valued by its owner for use or sale" (Porter 2002, 2). Perhaps not valued by its prior owner, waste can still be valuable for the waste management industry. Like waste, value is a contextual and informational concept, as demonstrated by objects that become valuable antiquities after being discarded as rubbish (Thompson 1979).

Often, the process of becoming waste implies a loss of information, a development that entails an ontological shift. Spanish philosopher Jose

Luis Pardo notes, "The process by which something is turned into trash can be described as a process of disqualification" (2006). Discarded objects lose their value by losing their characteristic properties and becoming part of an undifferentiated mass. The awe-inspiring aggregation of matter discovered in a hoarder's apartment or unloaded from a thirty-ton garbage truck renders the individual components invisible.

But producing waste also generates new information by sorting and separating trash from non-trash (Strasser 1999, 5). At the same time, sorting can reverse the process of becoming waste, restoring hidden information and value. This is the value proposition of a material recovery facility (MRF, pronounced "Murf"), in which co-mingled materials collected from curbside recycling are separated by composition and grade. Discarded plastic, paper, and glass have different commodity values based primarily on their levels of purity, which reflect how well the materials have been cleaned and separated using processes that often require significant resources of water and energy.

One might associate waste with noise rather than information, with entropy rather than order. Once inside the waste stream, however, waste is no longer "matter out of place," but subject to the regimes of diverse classifications. Not surprisingly, the early study of waste management by Rudolf Hering and Samuel Greeley began with a comprehensive categorization of municipal refuse (Hering and Greeley 1921). The European Union defines different kinds of materials in a regularly updated "List of Waste," and the U.S. Environmental Protection Agency (EPA) maintains similar lists for different kinds of hazardous wastes. The inclusion of a substance into such lists has far-reaching consequences—defining waste is a highly contested process.

Waste is a globally circulated commodity, the material basis of long-term service contracts, a calculated risk that can "bite back" long after its disposal. Waste systems cannot be separated from systems of production, and notions of value cannot be seen in isolation: "without disuse there is no use, and without waste there is no value" (Gille 2012, 28). Whatever is found in one system bears the imprint of the other, and systems of waste can be instrumental for studying systems of production. Understanding waste as part of a hybrid relationship between culture and nature, as sociologist Gay Hawkins argues, offers a productive alternative to a dualistic opposition between human and nature in which waste can play only a destructive role (Hawkins 2005).

## The Information Problem of Waste Systems

Solid waste management is the quintessential municipal service, the lowest common denominator of local government in the United States. Often, it just means collection and dumping at a local transfer station. But it is the gateway into a larger system connected to many different issues of public health, production, economic development, and urbanization.

The waste system is a global operation, yet it resists mapping. Its global dimension, structure, and flows can only be estimated. The *Handbook of Solid Waste Management* complains that local governments, policy makers, and waste management professionals lack accurate information about waste systems, a deficit it attributes to insufficient monitoring, the absence of shared definitions of waste in laws and trade agreements, and flawed coordination between administrative systems (Kreith and Tchobanoglous 2002). International estimates are often made without empirical basis or declared methods (Hoornweg and Bhada-Tata 2012). In short, there is no shared information model for waste.

For data enthusiasts, a lack of information alone might justify urgency. But many complex systems, such as the networks of informal recycling and waste picking described in part II of this book, are remarkably robust without any explicit monitoring or information infrastructures. It is therefore important to consider where, outside of pedestrian concerns about efficiency, insufficient data generates problems in the context of waste management. Disputes over waste policies and environmental justice are fought with data, statistical models, and maps describing situations that are difficult to observe. Here, a lack of information can lead to a lack of accountability, which can obscure pollution sources and cloak questionable practices. A lack of monitoring and enforcement can also undermine well-meaning policies, such as when incentives for recycling open loopholes that allow exporting electronic scrap as functional machines for "reuse" or abandoning hazardous materials held in "temporary storage." Finally, data about waste composition and provenance are instrumental for recapturing value from the waste stream. A circular economy in which discarded materials become new products requires information to differentiate the confusing mess that is waste.

Like commodities, all waste is traded (Porter 2002). Waste is a local issue and a global industry at the same time. Many materials such as mixed plastics are exported from the United States to recycling facilities in China—backhauled by the same vessels that delivered goods to the United States. In their reach and complexity, waste systems resemble global supply chains of production and distribution, with the significant difference that information tends to diminish during waste disposal. While manufacturers, retailers, and consumers can track goods and resources through supply and delivery chains, no such options exist for waste. Cities keep track of recycling rates, but the exported quantities and destinations go unreported (MacBride 2012, 181).

An engineering joke goes, "Practice is when everything works, but nobody knows why," meaning that complex systems often are not fully understood even by their managers and engineers. I have spoken to solid waste managers in local governments who did not know what happens to the city's trash beyond the first transfer station, which is typically where information reported by haulers ends. As waste and recyclable materials move through the waste system, it is not uncommon for them to change owners multiple times with little or no information exchanged or preserved in the process. It has been estimated that up to 80 percent of electronic waste ends up in developing countries despite various export bans (Lepawsky and Mcnabb 2010).

One might argue that global supply chains also suffer from information deficits. Consumers rarely know where product components have been sourced. Even manufacturers struggle with their end-to-end systems. Sourcemap, a Boston-based startup, helps companies investigate and visualize their supply chains. But this merely underlines the fact that real-time data about the movement of goods is crucial to the success of these operations, a requirement that has resulted in the standard use of bar codes and radio-frequency identification (RFID) chips to tag parts and packages.

For waste systems, the incentives are structured differently. Knowledge of what happens downstream is a question of public interest rather than a source of economic advantage. Gathering and exchanging information about the waste system is more difficult than in commercial supply-chain management or in other infrastructural systems such as power and water. Yet good governance of the waste system requires reliable information to evaluate whether recycling programs have their intended benefits, whether questionable disposal practices have been "greenwashed" by labeling them as recycling, or whether environmental crimes have been committed.

## The Agenda of This Book

As phones, dog collars, car keys, fitness wristbands, and a multitude of everyday objects become able to locate themselves and, to much concern, report their whereabouts to unknown recipients in a vast global network, a research question beckons. Can these gadgets partaking in the "Internet of Things" be leveraged to learn what happens inside the waste system? Emerging social practices concerned with tracking waste, documenting pollution, and producing environmental data beg another set of questions: Which roles do such civic technology initiatives and practices of appropriation and hacking play in the governance of public services? How are questions of accountability negotiated, and which roles do data representations and the design of mediating technologies play in this context?

When focusing on the larger goal of understanding the governance of waste systems, it is tempting to take technical methods, data formats, and protocols for granted. However, just as the different actors have their own ways of reading and representing waste systems, these technical details determine what we see when we try to read waste systems through the lens of technology. It is therefore helpful to examine how these technologies encode the material reality into symbolic languages and how they make the waste system legible through human and machine-readable languages, classifications, and social practices.

## What Is Information?

Until now, I used the term "information" interchangeably in a physical, abstract, symbolic, and material sense. Some clarifications are in order. Anthropologist Gregory Bateson defines a unit of "information" as "any difference which makes a difference in some later event" (Bateson 1972, 386). Although opinions might diverge about how "making a difference" should be interpreted, Bateson's definition is both broad and to the point, defining information by what it *does*, not by what it *represents*. It does not imply an omniscient, objective observer. It works from a subjective perspective in which it does not really matter whether the immediate observer is a human or an image sensor.

In the perspective of natural sciences, information is often treated as equivalent to a physical property, expressed in the "it from bit" hypothesis stating that all physical phenomena can ultimately be reduced to questions of information theory (Wheeler 1990, 311). Such a realist conceptualization of information is less popular in the social sciences and humanities, which mostly deal with information constructed by humans and subject to semantic ambiguities. Bateson's expression "makes a difference" subtly implies this constructed nature.

Information philosopher Luciano Floridi defines "information" as *meaningful data*, whereas a *datum* refers to a single difference, a lack of uniformity in a given context (2011, 85). This could be a difference in electric current, a microscopic bump in the groove of a vinyl record, or the difference between the symbols "a" and "ä." While this definition of "datum" sounds very similar to the earlier definition of "information," it implies that a datum is not necessarily meaningful, it does not always "make a difference." A valid and well-formed data set can be the result of a random process—meaningless.

How the processes and flows of waste systems are captured as data is a central concern in this book. I will use "data" to describe *a set of systematic observations that have been symbolically encoded and stored in material form*. This working definition implies that data are necessarily constructed through several steps. In order to collect a data set describing a waste stream, a method of observation has to be devised (such as sorting a truckload of garbage), a symbolic system has to be formulated for its representation (for example, a taxonomy of materials), and a method has to be chosen for encoding the observations into symbols. Finally, the observation has to be stored in a physical form. At each step of this process, decisions have to be made and often later questioned, renegotiated, and revised.

Because of the need to interpret data, visual theorist Johanna Drucker has argued that the Latin word *datum*,[1] meaning "the given," should be replaced by the active form *captum*, which is Latin for "the taken." She argues that data do not exist before they are parametrized, but are "constructed as an interpretation of the phenomenal world, not inherent in it" (Drucker 2011). Sharing Drucker's concern that the term "data" downplays all assumptions, decisions, and actions involved in their construction, geographers Rob Kitchin and Martin Dodge have rigorously used the term "capta" throughout a whole book, introducing terms such as "captabase" in the process. They define "capta" with a more realist flavor as "those units of data that have been selected and harvested from the sum of all potential data. … with respect to a person, data is everything that it is possible to know about that person, capta is what is selectively captured through measurement" (Kitchin and Dodge 2011, 261).

Although, for clarity's sake, I do not go as far as to adopt the "capta" terminology, when I describe "waste" as "information," I do not construe information as abstract, but as materially embodied. An item in the waste stream bears material traces of many social, cultural, technical, and political processes that can be scrutinized, whereas the concept of waste remains vague and ambiguous. Treating waste as information means following the heterogeneous network of connections in which a piece of garbage is embedded.

## An Iceberg Theory of Waste Systems

For something usually viewed as a problem, waste continues to fascinate. While there is a lack of information about waste systems, there is no shortage of perspectives and opinions about them. The two issues might be related. As Ernest Hemingway declared, the things that are left out are the most important parts of a narrative: "the dignity of movement of an iceberg is due to only one-eighth of it being above water" (1932, 192). Everyone has a view of the waste system, but as with the iceberg, the viewpoints are based on partial knowledge, and often our imagination is defined by what we do not see.

Iconic incidents in waste folklore have left a strong impression in our collective memory. In 1987 the trash barge *Mobro 4000* traveled to Central America and back in an unsuccessful attempt to offload 3,000 tons of New York City's garbage. Another popular myth claims melodramatically yet erroneously that the Fresh Kills Landfill on New York's Staten Island, closed in 2001, was the only human-made structure visible from space.[2] Detailed histories chronicle the environmental justice struggles from Love Canal to "cancer alley" along the Mississippi River. Many have heard reports about villages in Asia and Africa where electronics containing contaminants like lead are dismantled to reclaim elements like gold, since one ton of electronic waste contains 40 to 800 times the amount of this precious metal that is typically extracted from one ton of gold ore (Bleiwas and Kelly 2001).

While the apocalyptic imageries around Fresh Kills may exaggerate, other popular images underestimate the realities of waste management. For decades, the electronics industry was able to maintain the image of a clean industry despite the toxic legacies of semiconductor production (Gabrys 2013). This lack of information about the infrastructural processes involved in waste generation gives rise to imaginative but inaccurate theories, similar to the way anthropologist Willett Kempton demonstrated how homeowners' understandings of ubiquitous heat thermostats are inconsistent with technical realities (Kempton 1986).

How would the experiences of waste systems change if the public had more knowledge about their actual processes and geographies? What would we demand from municipalities? How would we express and support our doubts? How would the relationships and the interactions among citizens, governments, and other actors change? In short, how would this knowledge affect the governance of these systems?

Contemporary waste systems have been shaped by many different actors who use their own representational tools. Infrastructure governance and controversies are enacted through charts and tables, site maps and hydrological models, news photographs, and protests. Historian Martin Melosi describes how epidemiologists, engineers, citizens, and activists have approached waste as an issue of public health, an engineering problem, an aesthetic nuisance, or a manifestation of social injustice (Melosi 2004). As the range of these problems attests, the role of sanitation for modern urban planning can hardly be overstated. The overcrowded industrial cities of the nineteenth century were frequently struck by epidemics such as cholera and typhus. Ironically, a key driver of public health reform was a scientific misconception. The "miasma theory" that epidemics spread through contaminated air affecting rich and poor alike helped to establish consensus that these crises could be addressed only at the government level (Tarr 1996, 209).

As Melosi explains, the first to shape the waste systems were public health reformers such as Edwin Chadwick, whose descriptions of the unsanitary conditions among the urban poor set the basis for municipal sewer and garbage collection systems. Civil engineers saw public health reform as a technical and logistic challenge for building citywide sanitation infrastructures. With the City Beautiful Movement, affluent citizens gained influence over urban planning, perceiving waste primarily as an aesthetic problem that negatively affects moral sentiments and property values. Important for society but noxious for neighbors, waste facilities frequently became locally unwanted land uses (LULUs) that generated siting disputes and resulted in waste following the proverbial path of least resistance to marginalized communities. Toxic pollution in these neighborhoods gave rise to the environmental justice movement, which shed light on the politics of waste and contrasted with the goals and concerns of affluent environmental conservationists (Pellow 2004; Bullard 2000).

Even within the waste management community, one finds vastly different perspectives and agendas for composting, recycling, landfilling, waste-to-energy, or zero-waste. Not only do these approaches perceive waste as different kinds of problems, they use different toolsets for conceptualizing, observing, and representing the system. Epidemiologists, for instance, are concerned with spatial distributions of disease and medical pathways while engineers look at material flows and system performance. All representations share a common purpose in modeling the system as a coherent whole that can be investigated, manipulated, and optimized (Peattie 1987, 6). Each model defines inputs and outputs such as cases of disease and pollutants, or material flows and system capacities. Each input can be tweaked to achieve different outcomes that serve as a basis for decisions. By allowing different outcomes, each prescribes a specific perspective that offers a partial view of the system.

## The Shared Language of Location-Based Technology

Among all of these diverging perspectives, can there be a shared representation? Opening the Global Positioning System (GPS) as a public resource for civilian use has given rise to a large industry of location-based services that has played a substantial role in facilitating the global economy.

Within certain limits, GPS offers a shared language for investigating waste systems expressed in geographical coordinates, shared data formats, and tools for collaborative data collection, analysis, and visualization. Abstract and reductive, these generic representations offer a mere partial perspective that points to issues, stories, and experiences not included in the data. Nevertheless, the data formats and technical protocols are highly mobile, allowing transitions between different scales, contexts, and domains of knowledge.

Within the larger context of such *civic media technologies* (Gordon and Mihailidis 2016), this book focuses on geolocalization for reading waste systems through their spatial structures, temporal processes, participant connections, and system governance. Facilitated by smartphones and online platforms, civic technologies fall within a participatory culture that engages citizens in issues of public interest. Proponents of civic technologies envision citizens gathering information, tracking spatial processes, and visualizing complex systems in their roles as watchdogs, community organizers, DIY hackers, resource stewards, and expert amateurs (Ratto, Boler, and Deibert 2014; Kuznetsov and Paulos 2010).

Participatory sensing has been used to monitor everything from urban noise to radioactive pollution (Bonner 2012; Maisonneuve et al. 2009). Community-based efforts to collect geographical information have aided humanitarian responses to crises (Meier and Leaning 2009). Civic technologies include initiatives to offer government data in machine-readable formats for public access, allowing for public scrutiny of governmental processes and facilitating new products and services that utilize these data (Lathrop and Ruma 2010). By improving access to local government, civic technologies are assumed to make cities more responsive and accountable, engaging citizens in an ongoing conversation and collaboration with public servants and the public.

These positive effects, however, are often proclaimed rather than studied. Critics of civic technologies caution of inequalities in access and representation, threats to privacy and public anonymity, quality issues of data collected by self-selected volunteers, undue simplifications of complex social issues, or hidden agendas masked by narratives of participation (Jensen and Winthereik 2013; Boyd and Crawford 2012; Morozov 2014; MacKinnon 2012; Toyama 2015; Cooke and Kothari 2001).

Civic technologies often frame urban issues as a problem of participation and information exchange. They are grounded in a belief that the city can be improved by enabling all actors to talk to each other, and by making these interactions fast and effortless. They introduce new representations of the city using the languages of data, interactive interfaces, and dynamic maps that show fine-grained processes in real time. The role of design in mediating this interaction between systems and individuals is generally underappreciated. Interfaces play a central role in facilitating, shaping, and constraining interactions. A critical examination of civic technologies requires a close look at how interfaces and visual representations influence how information is collected, how meaning is constructed, and how action is taken.

## Infrastructure Legibility

While reading waste systems has not been a concern in urban studies, the notion that the city can be read like a text has a long history in urban planning and architecture. It has been applied mostly to city morphology—the shape of plazas, street fronts, and coastlines, as well as the topology of the road network. This book expands the concept of urban legibility to the realm of infrastructure and its governance. We read urban systems not only through our senses and experience, but also through public data repositories, visualizations, real-time data from sensor networks, and the traces left by other users and their actions. It is easy to forget that the experience of a system is shaped by the design of all these things, even if the experience of infrastructure is never fully determined or exhausted in design (Edwards et al. 2007, 28).

Throughout this book, I explore the connection between design and governance. I argue that design is in many ways a form of governance in how it shapes and regulates behavior, interactions, and conversations. Conversely, the processes of governance have many similarities to design

processes, often occurring as a series of incremental revisions and negotiations that political scientist Charles Lindblom called "muddling through"
(Lindblom 1959).

My central notion is that infrastructure governance is enacted through
the representations of the infrastructural system, and these representations
result from the efforts of different stakeholders to make the system legible
from their own perspectives and interests.

Many design choices shape the legibility of the waste system, starting
with its physical interfaces. New York City's waste bins come in many
different shapes and colors with no single visual language to designate
different material streams. In the streets, one can find separate bins for
beverage containers, newspapers, and residual waste. Bins in the subway,
however, accept all waste materials. Such inconsistencies send mixed messages even if governments want to encourage participation in recycling
programs.

Design decisions also determine how discarded materials are categorized as waste or recyclables. These decisions shape collection routes and
maintenance protocols that are explicated in service contracts and performance indicators. They determine how data are collected, which data sets
are accessible to the public, how they are represented, and how one can
gain access to them. Infrastructure legibility is not exclusively a concern of
service providers, however. Citizens, advocates, and activists also attempt
to make systems legible and to use system representations for their own
purposes.

Practices of making systems legible may include textual, visual, and
performative means. A substantial part of such practices is connected to
questions of accountability. Public restrooms often include a written journal of cleaning times, thus making the maintenance process legible and
holding the cleaning company accountable (Zinnbauer 2012). A constantly
overflowing waste bin indicates the need for a response that either reduces
waste generation or increases collection intervals.

Such disturbances can be an effective form of civic protest. When residents of a marginalized neighborhood in the Mexican city of Oaxaca
blocked access to the adjacent problem-ridden landfill, they caused trash
to pile up in the streets (Moore 2008). The performative display of their
protest and its consequences underscores that public policy has an underappreciated aesthetic dimension. The waste system may be defined by organizational and regulatory structures, but it is their material and tangible
consequences that determine how the system is perceived.

## The Design of Infrastructural Systems

When we hear the term "design," we might first think of the process of giving shape to a physical object, such as the iconic form of Eero Saarinen's TWA Flight Center at the John F. Kennedy International Airport in New York. The architect of an airport has to organize diverse functions in space and make them legible to travelers and employees. Physical design regulates how people check in, pass through security, and are monitored throughout the process. In this narrow sense, design is about organizing functions into a consistent, ideally delightful shape that "speaks" to its potential users. In other words, the design of an airport regulates user behavior, either subtly or, in the case of holding cells and other hidden security facilities, more forcefully.

Complex structures such as airports, however, often require additional measures to make sure that travelers do not get lost. *Information design* is concerned with organizing information, making it available where needed in a visual language that is both accessible and understandable. In the view of information architect Richard Saul Wurman, successful design facilitates understanding (Wurman 2000, 94).

Travelers increasingly check in using their smartphones, which have become part of the socio-technical system that regulates the physical space of the airport as much as its architectural features do (Dodge and Kitchin 2004). The coordination of the travelers' actions in physical and informational space is the domain of *interaction design*, which is concerned with how users engage with each other through a system. Websites and apps for booking and checking into flights require access to a technical layer of protocols, such as the proprietary AviNet Data Network Service or the federal Advanced Passenger Information System, which again are subject to design decisions and have to be coordinated between private and public stakeholders.

These protocols are not autonomous; they require the human action of traffic controllers or call-center operators. How customer representatives interact with travelers is a facet of *service design*, a discipline concerned with how services involving complex systems of people, interfaces, and technologies are organized and experienced. This practice includes specialized language and categorizations, such as the terminology the aviation industry uses to describe irate customers (Bowker and Star 1999, 37). As the *conversation designers* Hugh Dubberly and Paul Pangaro argue: "An organization is its language. Narrowing language increases efficiency. Narrowing language also increases ignorance. To regenerate, an organization creates a new language" (Shamiyeh 2014, 363).

Physical shape, the organization of information, people's behaviors, standards and protocols, languages and categorizations. All of these aspects are important for how we experience a trip. If one aspect is neglected, the whole system might fail. Applied to complex socio-technical systems, the definition of "design" is necessarily broad and inclusive. Due to their scope, socio-technical systems can no longer have a single designer and the concept of a single type of "user" is no longer applicable. Design concerns a manifold communication process between numerous actors, which cybernetician Gordon Pask described as a conversation (Pask 1976).

The diverse number of actors raises questions of power, equity, and authority. *Participatory design* and *co-design* describe approaches that include users in the design process. Participatory design emerged from the political context of Scandinavian trade unions that demanded a voice in the computerization of workplaces and factories (Kensing and Blomberg 1998). The purpose of participatory design practices is not limited to practical outcomes, however. It also works as a research method for learning about a specific environment through reflection on the participatory design process.

Applied to waste systems, a comprehensive design approach connects systems of production with systems of disposal. This is the intention of extended producer responsibility (EPR) policies, which incentivize manufacturers to consider recyclability in the design of a product by making them pay for its recovery. As media scholar Jennifer Gabrys describes it in the context of electronic waste, design should consider the whole "career plan" of an object from production, use and reuse, to final dismantling (Gabrys 2013, 152). The technical aspects are probably easiest to address in such a scenario. The human aspects, including occupational health concerns and questions of dignity, are more complex and potentially raise more controversies.

The overarching role of design in socio-technical systems does not imply that everything always happens according to plan. Infrastructures are also spaces of improvisation, temporary solutions, ad-hoc repairs. In this context, the designers are no longer outside the system, and their involvement is not over once a system assumes operation. In fact, many systems take shape as they are used, and design issues frequently arise from their use (Norman and Stappers 2015).

Urban planning and development literature concerned with infrastructure often still ignores the close entanglement of people and artifacts in urban services and conceptualizes the city exclusively as a product of abstract social, political, and economic forces. There are, however, several schools of thought within sociology and the studies of science and technology that embrace the world of material artifacts and their active role in urban systems. These literatures include actor-oriented and interface-oriented approaches, which avoid abstractions of social forces and instead investigate individual interactions at the boundaries between groups or systems (Long 1989; Lewis 1993). Langdon Winner's work investigates how technology embodies power relationships and influences social arrangements (Winner 1980). Social construction of technology (SCOT) literature investigates the development of technologies and how they are shaped by social interests (Bijker et al. 1987). Actor Network Theory (ANT) literature avoids any categorical distinction between people and artifacts in terms of their capacity to affect things in the world (Callon and Latour 1981).

To illustrate the interchangeability between human and nonhuman actors, Latour compares three means of traffic control: a sign with a speed limit, a policeman at the curb, or a speed bump in the middle of the road. In his view, all three are equivalent means toward the same goal, regardless of whether they involve humans or objects to achieve this goal (Latour 1994).

In traditional planning theory, this might seem to be a provocative position, but it deeply resonates with the diverse practices of design, which are all concerned with how material artifacts shape their environment, how they affect behavior and communication, and how they are affected and changed in their use. In short, with the differences that material artifacts make in the world.

## Design and Legibility

All design practices share a concern for what should be hidden and what should be exposed. A prominent principle of a functionalist understanding of design is to make the system as unobtrusive as possible (Buchanan 1985). Computer scientist Mark Weiser asserted that computers had to become invisible to become ubiquitous. Comparing computation to the cultural technique of writing, he observes: "The most profound technologies are those that disappear. They weave themselves into the fabric of everyday life until they are indistinguishable from it" (Weiser 1991). Or as Bowker and Star note, "Good, usable systems disappear almost by definition. The easier they are to use, the harder they are to see" (1999, 33).

One strategy to "make a system disappear" is to hide its technical complexities and make its surface as seamless and consistent as possible. By minimizing the number of inputs and moving parts, a system can become more reliable and accessible. Such a paradigm of *seamless* design has drawbacks, however. Without the option for intervention in a simple interface, the user may not feel in control, especially if something does not go as expected. A seamless system can be difficult to adapt, upgrade, repair, or even diagnose in the case of a breakdown.

Instead of aiming for a homogeneous and monolithic system, *seamful design* emphasizes the seams (Chalmers and Galani 2004)—similar to ceramics repaired in the Japanese Kintsugi technique, which emphasizes the seams between the broken pieces. By deliberately exposing certain technical aspects, a system can become more adaptable and extensible, providing visible clues about its internal state. Provocatively, this could mean designing a system that is easily "hacked" through appropriation, study, and improvement by the user, accepting the possibility that parts of the system might break in the process (Galloway et al. 2004).

Given the prominent concern with invisibility, it seems contradictory that an equally important goal of design is to make objects and systems "talk" by exposing their functions to the user. In reference to this paradox, industrial designer Dieter Rams explains that design becomes unobtrusive by being informative about its functions rather than by evoking emotional response (Rams 1984). In Heidegger's terminology, designers who subscribe to the functionalist paradigm exemplified by Rams and others are concerned with producing artifacts that are "ready-to-hand," that exist in relation to other things, techniques, and actions, rather than "present-at-hand," a solitary state that becomes manifest when a tool becomes unusable (Heidegger 1927, 73).

How can this desirable state be achieved? Since the 1980s, the cognitive aspects of design have been inspired by J. J. Gibson's ecological approach to visual perception. Gibson's *affordances* describe the visible properties of the physical environment in relation to what they offer an animal, such as shelter in the shape of a cave (J. J. Gibson 1979). In Donald Norman's reading, a user constructs a *system image* of an object and its functions by interpreting its affordances. More than simple signs based on convention, affordances are possibilities for action. A metal bucket, for example, affords turning it over and stepping on it, making noise with it, or filling it with water. To learn how to use a system may involve clicking buttons to see what happens. Designers can therefore communicate by creating affordances that signify a system's state and its possible actions (Norman 2002, 188).

This is not to say that symbolic information is not important. Other than with simple acts like using a bucket, things rarely speak for themselves. "Intuitive" interfaces are more accurately called "clear, if previously understood," since they depend on assumptions of shared knowledge and experience. Because the heterogeneous infrastructures considered in this book encompass organizational and informational aspects as well as physical components, design in these contexts involves making a system legible by considering possibilities for action, questions of symbolic encoding, the organization of information, and associated human practices.

## How This Book Is Organized

Based on the concept of infrastructure legibility, this book investigates existing and emerging methods for reading waste systems and examines how they influence the governance of these systems. The three parts of this book look at the formal structure, informal practices, and the interactions between individuals and service providers. They address a common problem of how to gain information about what happens "inside" waste systems given the scarcity of data and a lack of incentives to collect and share them. The case studies represent three ways to address this question:

1. by using sensor technologies for tracking and observing waste flows,
2. through context-specific data initiatives driven by local actors,
3. by appropriating and analyzing existing data sources.

In the context of urban data initiatives, the opacity of the waste system serves as a counter-example to the widespread assumption that the world is drowning in data. The three cases are therefore not arbitrarily chosen. They act as exemplars for the three most important directions in current urban data research initiatives, which include the Internet of Things approaches that use networked sensors, context-sensitive data strategies tailored to a particular community and situation, and urban analytics methods that mine existing data repositories for information about the state of urban systems. Each of the three methods represents a different way of reading urban infrastructures, and each comes with its problems and limitations, addressed in the following three parts.

I undertook these studies as part of a team of researchers from the MIT Senseable City Lab directed by the architect Carlo Ratti. All three studies make use of location-based technologies such as GPS to map the geographies of waste transportation, the organizational arrangements of service providers, and the interactions between citizens and local governments. Each study looks at waste infrastructure through a different lens, including the external perspective following items through a waste system, the internal perspective of workers within a waste system, and a boundary perspective of citizens crossing into the system to engage with service providers.

### Part I: Legibility

Part I investigates how waste infrastructures are represented and made legible through existing monitoring systems, and how these representations have shaped the public discourse around waste systems. It compares two foundational conceptions of urban legibility proposed by Kevin Lynch (1960, 1984) and James C. Scott (1999). Although remarkably different, these conceptions provide complementary perspectives for understanding infrastructure legibility, capturing physical-sensory as well as abstract-informational dimensions in the design and the politics of urban systems. Based on this foundation, I develop a framework for investigating how infrastructure systems such as waste collection and recycling are made legible.

The Trash Track case study that my colleagues from the Senseable City Lab at MIT and I carried out in Seattle examined the geographies and topologies of companies and facilities involved in the waste system (Boustani et al. 2011). With the help of active location sensors, we mapped how items discarded by a typical household moved through the waste and recycling system, gathering data that is difficult to obtain using traditional methods. The approach prototyped in this study allows us to estimate the extent to which transporting waste diminishes the benefits of recycling, to support the investigation of environmental crimes, and to offer ways of monitoring a waste system from "the outside."

### Part II: Informality

Part II focuses on the challenges that developing economies face when moving from informal to formal systems for waste management. The tools of legibility in the process of formalization involve legal, technical, and social arrangements, many of which are driven by a need for accountability in assessing service contracts, policy options, environmental pollution, and public communications.

The Forager study described in part II looked at the spatial strategies, social organization, and tacit knowledge of waste pickers working for Brazilian recycling cooperatives in São Paulo and Recife. It investigated the informal processes these workers used to organize their collection routes and processes, exchange data, and interact with governments, companies, and the broader public. These cooperatives operate mostly on tacit knowledge, and the study examined ways to improve their data collection methods to benefit their dealings with corporations and municipalities without petrifying their dynamic, flexible organizations.

The result of this inquiry, however, is not necessarily an optimistic story of empowerment. Instead it highlights the limitations and dangers of "formalization by measurement" in which location-based technologies expose fragile, informal systems to more powerful actors.

## Part III: Participation

Part III looks at how infrastructure monitoring and participatory infrastructure governance are enacted through civic technologies such as citizen feedback systems. Urban infrastructures in many cities of the United States have become more visible and more participatory, for better or worse. Due to limited budgets, cities struggle to maintain urban infrastructure, and due to privatization of service provision, individuals are exposed to a more fragmented and unequal infrastructural landscape (Steve Graham and Marvin 2001). Various solutions are advertised by technologists. We are told that live data from sensor networks can make infrastructure more efficient, that performance metrics can make governments more effective, and that digitally mediated participation can make cities more civic and just.

To examine the role of participatory sensing in infrastructure governance, my third case study analyzes the design and use of mobile systems that allow citizens of Boston to report infrastructure failures such as potholes, broken streetlights, or garbage spills. It looks at how design of the various interfaces between citizens and government shapes their interaction, as well as how a service provider's self-representation affects its legibility. This part raises questions about the role of mediating technologies in governance models that are based on interactions between governments and constituents. It makes a case for scrutinizing the political nature of interfaces and argues that standards and common protocols are critical components of democratic discourse.

## Conclusion: A Case for Accountability-Oriented Design

The conclusion synthesizes the forms of legibility described in the book, extending the notion of infrastructure legibility beyond waste systems to address concerns about accountability, anonymity, and the role of the user in distributed systems. The conclusion makes a case for accountability-oriented principles that acknowledge the political responsibilities of design. Because interactions between individuals and system providers are shaped by interfaces, the designer often unknowingly assumes the role of regulator and moderator. I finish the book with a proposition for design principles that calls attention to hidden issues of governance and that situates the design and implementation process of interfaces, protocols, and platforms as a form of democratic discourse.

# I   Legibility

# Prologue to Part I: Tracing Waste Geographies

**Figure I.1**
Waste trackers reporting from a landfill in Arlington, Oregon. U.S. Geological Survey,
State USDA Farm Service Agency, reproduced under Google Maps fair-use policy.

Urban infrastructures are often characterized as invisible, but this is not
entirely true, since they equally capture our imagination. Their complexity
is only revealed in partial perspectives, since they consist of such diverse
elements as physical installations, the coordinated practices of workers,
organizational structures, laws, patents, and regulations.

Chapter 1 investigates tensions in the experience of infrastructure—
how the waste system is represented in the data collected by its diverse

monitoring mechanisms, and how these representations influence public conflicts around pollution and environmental justice.

The notion of "reading" reoccurs throughout this book. I argue that it is necessary to look at the data representations of infrastructure not merely as encoded facts that can be analyzed as such, but rather as traces that must be interpreted in the context of the material reality of the system and the specific methods of data collection. For example, the GPS location reported by a tracked object might seem like a clear indication of the object's presence at the reported location, but the causal interpretation of the underlying processes is complicated by many different factors.

The case study presented in chapter 2 examines an experiment to render waste infrastructure legible by observing the movement of different waste streams. The experiment attached location sensors to typical items of trash found in private households throughout the city of Seattle, including beverage containers, papers, shoes and textiles, toys, appliances, obsolete computers, and old cell phones. The goal was to map discarded items as they moved through the waste system and to identify companies, facilities, and logistic networks. From an analytic standpoint, the project aimed at collecting baseline information about waste transportation distances and destinations and investigating how predictably they move. While many aspects of waste logistics are regulated in service contracts and legislation, there is surprisingly little information about how waste actually moves, making it difficult, by traditional means, to address the question of how hauling distances can diminish the benefits of recycling.

The study is based on the MIT Trash Track project, which I was involved in during my time at the Senseable City Lab[1] from 2009 to 2013 (Senseable City Lab 2009). The lab is dedicated to studying the possibilities and issues of data and real-time sensing technologies for cities and the urban planning discipline. Early projects of the lab visualized urban activity through the aggregated digital exhaust of cell phone networks and other information infrastructures. The visualizations and data models developed in these projects offered an unfamiliar lens on urban activity through the proxy of mobile communication.

Compared to these projects drawing from existing, but previously unobtainable data sources, Trash Track was radically different, since it focused on a domain where virtually no data existed. The project originated from an exhibition proposal for the 2009 "Sentient City" exhibition curated by Mark Shepard at the New York Architectural League (Shepard 2011). Rex Britter, professor emeritus of engineering at Cambridge University and visiting researcher at MIT, suggested tracking garbage in the city to investigate what Bill Mitchell described as the "removal chain," a mirror image of the supply chain. The exhibition budget allowed only a small deployment of sensors in New York—not enough to do justice to the vast extent of the system under examination. Through partnerships with companies and public institutions, namely Waste Management Inc., Qualcomm, the city of Seattle, and the Seattle Public Library, it was possible to conduct a larger experiment on the West Coast of the United States. I joined Trash Track in September 2009 to lead the second part of the project after completing the New York exhibition: the deployment of approximately 2,500 waste sensors in Seattle, Washington.

Trash Track was an effort of many individuals. Overall, the project involved more than fifty people, including urban planners, environmental engineers, designers, software and hardware developers, student researchers, and lead volunteers. The deployed sensors reported waste movement over the period of five months, followed by data analysis. It took another two years until the results were published. The publications covered many aspects and disciplines: urban planning implications (Offenhuber et al. 2012), sensor development and pervasive computing (Boustani et al. 2011; Phithakkitnukoon et al. 2013), electronic waste management (Offenhuber, Wolf, and Ratti 2013), and the volunteers' perceptions of the waste system (Lee et al. 2014).

Through the involvement of volunteers, the project took on a life of its own. Tracking as a way of reading the city and connecting it to one's personal experience was a recurring theme throughout the experiment. As one participant noted on the Trash Track volunteer survey: "I hadn't thought about the trash having multiple stops between me and a landfill. I also realized I have no idea where my local landfill is" (Lee et al. 2014).

The volunteers joined the project for different reasons: many were interested in making waste systems more sustainable, while some saw it as a clandestine method to investigate the city's waste policies. But for many volunteers, the project had a profoundly personal significance, helping them to understand the consequences of their own decisions. In many cases, the participants brought objects of personal significance to follow their last route as trash in the waste stream.

The digital artifacts created by the trackers attached to discarded objects afford a way to experience the waste system through a particular lens. Part I of the book will examine the different ways of experiencing infrastructure through data representations and what role these representations play in public discourse and governance of these systems.

# 1 Visibility

We realize that involvement in this type of research is a departure from our traditional protest methodology. However, if we are to advance our struggle in the future, it will depend largely on the availability of timely and reliable information. We believe this data should be utilized by federal, state and municipal governments to prevent hazardous wastes from becoming an even greater national problem. No residential community, regardless of race, should be left defenseless in the midst of this mounting crisis.

—Benjamin F. Chavis Jr., preface, *Toxic Wastes and Race in the United States* (United Church of Christ 1987)

## Toxic Wastes and Race

In the first national study on environmental justice, *Toxic Wastes and Race in the United States*, of 1987, the United Church of Christ Commission for Racial Justice presented a statistical analysis that confirmed what had frequently been suspected: undesirable land uses such as hazardous waste landfills and processing and storage facilities are preferably sited in areas where disadvantaged populations live. The study found that among a number of socio-demographic variables, the relative size of African-American and Hispanic populations was the most significant predictor for the number of hazardous waste facilities in an area (United Church of Christ 1987).

To reach this conclusion, the study analyzed two data sources that had only recently become accessible to the public. The first was the Hazardous Waste Data Management System (HWDMS), a register of commercial hazardous waste facilities. The second was CERCLIS,[1] a directory of uncontrolled hazardous waste cleanup sites. Both of these databases are maintained by the U.S. Environmental Protection Agency (EPA), as mandated by federal laws that were enacted to regulate waste management after the toxic waste disasters of Bhopal and Love Canal.

HWDMS and CERCLIS are part of a larger information infrastructure designed to monitor the waste system on the national level. This system involves many actors and combines data generated by scientific studies, submitted by companies or members of civic society, and collected by local governments and federal enforcement agencies. Despite its size and heterogeneity, the picture of the waste system captured by this information infrastructure is far from complete. Many of its aspects, definitions, standards, and responsibilities are contradictory across administrative units, ambiguous, and unresolved. The information captured is the result of political struggles over scientific findings, definitions, and legal procedures. Yet the difficulties in capturing the processes and heterogeneous realities are not unique to the waste system; they are characteristic for reading and representing infrastructure in general.

While few members of the public are aware of information infrastructures such as HWDMS and CERCLIS,[2] such systems influence the public experience of infrastructure. They are used by governments, journalists, and advocacy groups to create representations that shape the shared perceptions of these systems. Since public opinion and media pressure have always been main drivers for environmental legislation, these perceptions can have consequences for the architecture of information infrastructures and monitoring systems shaped by legislation. Environmental monitoring and the experience of infrastructure are therefore closely connected.

To illuminate the contingencies and mechanisms of monitoring the waste system beyond the scope of the hazardous waste stream, it will be necessary to step back and look broadly at conceptualizations of infrastructure and the forces that shape the manifestations and practices "infrastructure" describes. To delineate what the "waste system" is and what it looks like is no trivial task, since the system presents itself as a hybrid arrangement that includes facilities and machines, work schedules and organizational arrangements, and legal and scientific reports. It involves the biochemists who investigate the threshold levels of chemical exposure as well as the public managers who define the terms of service contracts. The waste system as an analytical object shares many representational, phenomenological, and ontological challenges with other kinds of systems, which are subject to the growing field of infrastructure studies. As we will see, all readings and representations of the waste system are necessarily incomplete and based on a particular goal and interest. By introducing the framework of infrastructure legibility, I trace the connections between different data sources and the aspects of the waste system that they cover. In this respect,

**Figure 1.1**
Statistical Exhibits in the Municipal Parade by the Employees of the City of New York, May 17, 1913. "Many very large charts, curves and other statistical displays were mounted on wagons in such manner that interpretation was possible from either side of the street. The Health Department, in particular, made excellent use of graphic methods, showing in most convincing manner how the death rate is being reduced by modern methods of sanitation and nursing" (Brinton 1914, 342).

later in this chapter I will describe six different elements of making infrastructures legible.

## Experiencing Infrastructure

With about fifty thousand miles of asphalt and concrete, the U.S. Interstate System provides no single point where one can take it all in. From space, the delicate structures disappear in relation to the area they cover. From the ground, the same stretch of freeway looks very different to a speeding motorist and to a pedestrian trying to cross it. The observed qualities of a system depend on perspective and scale. What may present itself as stable from close range can become fragile in a larger view of space and time (Edwards 2003).

Although the term "infrastructure" implies something submerged below surfaces or concealed behind walls, urban systems are right in front of our eyes, hidden because they are too large, too small, or too complex to be seen in their entirety. Taking Paris as their example, Latour and Hermant argue that the city is invisible because it can never be seen in its totality (2004). It can be perceived only through partial perspectives and representational systems such as the street name register, control room displays, or composite satellite images.

Representations such as diagrams, maps, or data sets are central to reading infrastructures, but they rarely fit together neatly in a coherent and complete image. Each representational mode emphasizes a single aspect and obscures others (Peattie 1987). A lack of representation, both in the visual and political sense of the term, can render a system invisible, including the people who maintain it.

A system may be described as invisible because relevant information does not exist or its governors lack oversight. Older systems contain many obsolete components whose original purpose has been forgotten, and which therefore cannot be safely removed (Hughes 2004, 79). Sometimes information is deliberately withheld as with the minute details of the electric grid, the trade secrets protected by intellectual property laws, or the algorithms used to generate credit reports. This has social and political implications. An invisible system does not raise questions.

Invisibility can also be the result of too much familiarity, habituation to infrastructural systems that has left the user "sleepwalking" in their midst (Winner 2014; Star and Ruhleder 1994). Users who trust a system's functions have no interest in its complexities, treating it as a black box reduced to inputs and outputs (Latour 1999, 305). Visibility and invisibility depend on the modes of experience. A flight of stairs might be an infrastructure for one person and a barrier for another (Star 1999).

Of course, infrastructures can also be highly visible, even monumental. Apart from landmarks such as bridges or radio towers, even landfills and transfer stations can offer everyday spectacles that rival those of airports or train stations. Yet projects initially received as spectacle often eventually slip into banality (Steve Graham and Marvin 2001, 21). The impression left by the first electric street lighting systems, described by David Nye as the "electrical sublime," and the enthusiasm inspired by public sewer systems in the late nineteenth century are difficult to imagine today (Nye 1992; Tarr 1996).

Sometimes this process of banalization can be reversed. One of Vienna's more peculiar tourist destinations is an operational waste-to-energy plant, which handles the residual waste of the Austrian capital. After a fire had damaged the incinerator and public opposition mounted against restoring the facility, Mayor Helmut Zilk commissioned a prominent environmentalist artist to redesign the building as a flamboyant architectural sculpture. However blatant this tactic of visual seduction might seem, it succeeded in launching a sophisticated public debate about the role of incineration in a country with no more space for landfills. Geoffrey Bowker describes this phenomenon in which previously invisible infrastructure comes into sudden scrutiny as infrastructural inversion, a process that reveals the core assumptions about a system's underpinnings (1994, 11).

The black boxes of infrastructure are rarely fully closed, nor can they be kept tightly shut. An otherwise invisible system becomes visible when it fails (Star 1999; Graham 2009). Infrastructure breakdowns do not necessarily have to be catastrophic—they can be instructive by revealing a previously unknown aspect. Even in the absence of breakdowns, daily life involves struggles with infrastructure as we navigate a multitude of systems and encounter the obstacles they present.

Urban systems oscillate between visibility and invisibility, move between center and periphery of attention. Despite our sometimes difficult, sometimes blissfully unaware relationship with urban systems, we are not simply subjected to invisibility. Our daily interactions with urban systems and their representations offer clues from which we construct models of how these systems work. Some aspects of these ad-hoc explanations might be accurate, others not. Infrastructural representations reinforce each other and cannot be compartmentalized to a specific group, practice, or purpose. The waste system is tightly interwoven with other systems such as transportation infrastructure, the electric grid, or the contracts of public employees, for example. Each component is present in multiple systems at the same time.

## Definitions of Infrastructure

As is true of experiencing infrastructure, the definition of infrastructure involves multiple perspectives. Often, attempts to define "infrastructure" resort to open-ended lists of examples to clarify what kinds of systems are meant. The dictionary glosses "infrastructure" as "the subordinate parts of an undertaking; substructure, foundation; specifically the permanent installations forming a basis for military operations, as airfields, naval bases, training establishments, etc." (OED 2012).

Definitions of "infrastructure" vary by discipline and profession. Urban planners and engineers distinguish between the *hard* infrastructure of physical structures such as harbors or sewers and the *soft* infrastructure of organizational structures such as healthcare or education. This clean separation is misleading because even the physical components require complex social arrangements for their development, maintenance, and use. All of these arrangements continuously evolve: new recycling procedures are introduced, new contracts closed, and new partnerships entered. This is to some extent reflected in the distinction between *infrastructure* and *utility* and *service*, in which infrastructure refers to the physical structures, operated and maintained by utility companies, in order to provide services.

### Conceptualizations from the Perspectives of Builders and Users

While most definitions of "infrastructure" have a disciplinary viewpoint, the lived reality of urban systems transcends disciplines and professions. Urban systems have historical, social, scientific, and political dimensions, touching almost every aspect of life. Historian Thomas P. Hughes speaks broadly of Large Technological Systems (LTS), which he characterizes as problem-solving, complex, and messy, both socially constructed and society shaping (Hughes 1987, 51). LTS are complex socio-technical organizations with many stakeholders and encompass physical, theoretical, and legal components (Mayntz and Hughes 1988). Large technological systems evolve, mature, and change over time, just as their components interact and co-evolve. If a new technology or a law affects one part, all other parts adapt accordingly (Hughes 1987). Despite its broad scope, the LTS model is not complete. It conceptualizes infrastructure from the position of system builders. The users of a system play a marginal role in this perspective.

In contrast to this "view from above," anthropologists Susan Star and Karen Ruhleder describe infrastructure from the user perspective as an emergent activity rather than a physical structure, "fundamentally and always a relation, never a thing" (Star and Ruhleder 1994, 253). In this view, the qualities of a system are not predetermined, but emerge from the practices of its users. Star later defines "infrastructure" through a set of properties that characterize its relationship to the user. In her characterization, infrastructure becomes transparent in use and visible upon breakdown, is shaped by conventions that have to be learned, and is "fixed in modular increments, not all at once or globally" (Star 1999). A special case in this user-centric view are systems that are entirely built and maintained by their users (Egyedi and Mehos 2012). The maintenance of such "[i]nverse infrastructures" depends on coordinated and voluntary tasks by users, which makes these systems difficult to sustain and relatively rare in the urban landscape. Nevertheless, some examples exist, such as the volunteer-driven collection of waste paper in parts of the Netherlands (De Jong and Mulder 2012).

It might be tempting to place LTS and inverse infrastructures in rhetoric opposition as top-down vs. bottom-up systems. Such binary categories are misleading, since centrally managed large technological systems and self-organized inverse infrastructures share many similarities. Just as centrally planned systems depend on bottom-up engagement to work, self-organized systems often involve centralized governance structures. The roles of builders, regulators, and users have become increasingly blurry, as I further discuss in part III of this book.

### Economic Conceptualizations and Their Consequences for Infrastructure Monitoring

Notwithstanding the technical, organizational, and social aspects of infrastructure, the reality of urban systems is to a large part determined by economic forces. Infrastructure development requires extensive investments, and questions of financing often dwarf technical and organizational problems. Because public utilities increasingly are owned by investment banks and pension funds, economic geographer Morag Torrance argues that "infrastructure" has become first and foremost defined by its nature as a financial product rather than by its technical properties and social purposes (Torrance 2009, 807).

From the economic perspective, the question of who builds and who uses infrastructure is reframed into a question of supply and demand. From the supply side angle, economist Remy Prud'homme defines "infrastructure" as a capital good for the purpose of service provision (2005). He points out that investments in these goods tend to be lumpy rather than incremental, requiring large sums upfront for delayed returns: an almost-finished bridge offers no practical value. As a result, infrastructure development is slow to respond to increasing service demands, and it becomes advantageous to over-size capacity in the first place rather than risk more expensive upgrades later.

While a public-sector service provider can cover the lumpy initial investments through bonds or taxes, a private-sector service provider will go the route of fine-grained measurement and billing of individual service consumption—a task that requires sophisticated information infrastructures. The question of whether service consumption can be measured can determine whether a service eventually becomes a private or a public good, which will be discussed in further detail later in this book.

In economic terms, a good is public or private depending on rivalry and excludability, factors that address consumption and access, respectively (table 1.1). The rivalry factor concerns whether a resource diminishes when used; excludability refers to which groups can consume a resource. If the cost of enforcing exclusion is higher than the cost of public access for the service provider, a service becomes a public good. For example, the use of private waste bins is excludable, while the use of public waste bins is not. However, no infrastructure service is purely nonrivalrous: waste bin space along with the capacity of roads and the availability of water and clean air are limited, and they should therefore be more appropriately understood as common goods.

Calling attention to the demand side of infrastructure, economist Brett Frischmann argues that infrastructure should be viewed as a common good because the social returns of investment exceed private returns (2012). Combining aspects of both public and private goods, common goods introduce their own rationale for monitoring infrastructures, which is focused on the management of limited resources rather than the recuperation of initial investments. As famously demonstrated by economist Elinor Ostrom, even user-managed, informal infrastructures involve complex governance structures for collective monitoring, information sharing, and enforcement to prevent overconsumption of the common good (Ostrom 1990).

## Infrastructure Awareness, Accountability, and Governance

Due to their multifaceted nature, infrastructures are difficult to read and represent, and under normal conditions, users might not be aware of their presence. Yet the governance of infrastructures such as the waste system relies precisely on this kind of public awareness: the capacity to read systems and act upon this information.

A successful regulatory instrument for hazardous waste in the United States, the Toxics Release Inventory (TRI) program does little more than collect data and make it public. It requires firms to report chemical discharges,

**Table 1.1**
Public and private goods from an economic perspective

|               | Excludable    | Nonexcludable |
| ------------- | ------------- | ------------- |
| Rivalrous     | Private goods | Common goods  |
| Nonrivalrous  | Club goods    | Public goods  |

and it makes this information available in a public database. An annual TRI report tracks the amount of chemicals that a facility discharges as well as whether the chemicals are incinerated, injected into the ground, or transported to other facilities. The rationale behind this regulatory strategy is that firms forced to report toxic releases will become more mindful of their environmental impact and that the data collected and publicized can become a tool for investigative journalists and advocates to create social pressure. Over half of all reporting facilities have made changes to their operations as a result of their reporting, and emissions have dropped significantly during the first years of the program's operation (Guerrero 1991; J. Hamilton 2005).

The TRI database has become an essential proxy for estimating the amount of toxins released over time. Yet the effectiveness of TRI for spurring environmental action depends not only on its accuracy and conclusiveness, but more importantly, also on the possibility to interrelate, cross-reference, and augment it with other data sources. In the framework of environmental information systems, TRI is merely a part of a larger information infrastructure that combines data of multiple types using multiple sources and collection mechanisms.

## Community Right to Know

The TRI program had its origin in the 1986 Emergency Planning and Community Right to Know Act (EPCRA), which mandates that communities need to be informed about pollution they are potentially exposed to and the harm these exposures might cause. Its *right to know* provision spells out the duty of regulators to generate information, to store it, and to make it accessible to the public upon request. In the case of imminent harm, communities must be actively informed. Besides information about discharges, emissions, and facility locations, these provisions include the collection and publication of scientific data about chemical substances and their health effects.

Such data were collected from epidemiological and medical studies during the first stages of a different environmental program, the Resource Conservation and Recovery Act (RCRA), which is the main federal law concerned with waste management (U.S. Congress 1976; U.S. Congress 1984). As with environmental programs such as the Clean Air Act and the Clean Water Act, the first years of implementing RCRA were dedicated to collecting scientific data to determine which wastes are hazardous. This turned out to be an enormous and contentious task. In a 1987 report, the Government Accountability Office (GAO) noted that despite ten years of data collection, the "EPA does not know if it has identified 90 percent of the potentially hazardous wastes or only 10 percent."

Besides identifying hazardous substances, RCRA regulates the processes of hazardous and solid waste management. RCRA's "cradle to grave" approach regulates the generation, storage, transport, treatment, and disposal of hazardous wastes, authorizing the EPA to monitor compliance of permitted facilities by conducting site audits and environmental tests. The 1984 amendments to RCRA and EPCRA introduced a stricter regulation that required permitting and reporting for facilities that treat, store, or dispose of hazardous wastes, laying the groundwork for establishing the TRI database three years later.

### Citizens as Watchdogs

RCRA, however, is not responsible for the toxic legacies of already polluted sites. Known as the Superfund program, the Comprehensive Environmental Response, Compensation, and Liability Act (CERCLA) was enacted in 1980 in response to the Love Canal disaster, and addressed the cleanup of abandoned or uncontrolled toxic waste sites. Funded by taxes on the oil and chemical industries, the law imposes strict liability, which means that firms are held fully liable for their waste management subcontractors' actions, even if the firms are not aware of their conduct. This is intended to force businesses to vet and monitor their subcontractors. The identification of cleanup sites depends on information submitted by members of civic society, just as the Love Canal incident was brought to public attention through the coordinated action of local residents.

While the scope of EPCRA[3] is limited to information collected by a federal agency, RCRA and CERCLA include other provisions that give a more active role to citizens. Both programs include a citizen lawsuit provision, which allows any person to bring a suit against suspected polluters that "may present an imminent and substantial endangerment to health or the environment" (U.S. Congress 1984). The success of citizen lawsuits, however, has so far been limited. The interpretations of terms such as "imminent" and "harm" have been subject to lengthy legal battles. Scientific uncertainty about exposures and health effects, missing or unreliable data, and different legal interpretations have turned citizen lawsuits into complex endeavors with uncertain prospects.

### The Changing Role of Civic Society in Waste Policy

Early U.S. environmental legislation such as the Clean Air Act defined environmental regulation and enforcement clearly as a federal responsibility. In the later programs, including the Clean Water Act and RCRA, the distribution of responsibilities has become more complex. Under RCRA, nonhazardous and municipal solid waste are subject to regulation by states while hazardous waste remains a federal responsibility. In the wake of President Reagan's regionalist agenda, more responsibilities shifted to state agencies, which were encouraged to introduce their own approaches, standards, and enforcement mechanisms. The transition from the early command-and-control approach strictly enforcing health-based standards to the quest for "socially optimal levels of pollution" designed to balance harms and costs introduced new data-driven assessment approaches such as cost–benefit analysis and quantitative risk assessment.

The larger number of stakeholders involved in defining and negotiating environmental policy created more space for the participation of civic society. At the same time, a greater number of chemical substances has made the nature of harm harder to assess. Endocrine-disrupting effects of plastics and the interaction between different pollutants are still subject to scientific uncertainty, mostly due to the vast scope of the phenomenon and the limited resources to conduct studies.

Owing to the messy ambiguities that result from the multistakeholder process, RCRA has the reputation among legal experts of being the most complex and arcane among the environmental laws (Garrett and Association 2004). Supported by scientific uncertainty, environmental policy is an overtly political process for negotiating different approaches, from the precautionary and equity oriented to the more utilitarian. The gradual withdrawal of federal agencies from environmental issues has placed a larger burden on nongovernmental agencies (NGOs), watchdog organizations, and civic society to open the black boxes of infrastructural systems and get involved in governance.

## Alternative Practices of Reading the Waste System

Data collected by public authorities about the waste system are limited in several ways. TRI forms are self-reported, and therefore unlikely to include information about potential violations. The 690 chemicals that have to be reported do not represent all relevant pollutants. In 1991, the GAO estimated that 95 percent of the actual toxic releases are not covered in the database (Guerrero 1991). Subsequent expansions of the TRI are hardly keeping pace with the growing numbers of substances, pathways, and their potential interconnections.

RCRA's monitoring and tracking requirements are limited to certain hazardous wastes. Municipal solid waste is exempt even if it contains toxic materials. Certain hazardous wastes discarded by households, such as cathode-ray tubes (CRTs) of monitors and TV sets, fluorescent light bulbs, and printer cartridges, are differently regulated in different states. A substantial amount of waste goes unreported in the national municipal solid waste (MSW) statistics due to an absence of shared definitions and a lack of quality data. In the context of RCRA, the roles of federal, state, and local governments in waste regulation are not always clear, which complicates the enforcement of violations (Kreith and Tchobanoglous 2002). Transportation of MSW and nonhazardous wastes is not monitored. Therefore, data about waste movement within a state or between state and national boundaries is largely unavailable. The trade and tariff classification system does not capture information about exported and imported waste.

Alternative sources of information are scarce. Firms lack incentives to collect data beyond the scope of strict liability since knowledge of what happens downstream offers no benefits to a company. Local and regional governments also lack incentives to play a stronger role in pollution control since they often do not want to take action against businesses that generate taxes and employment. Finally, regulatory agencies often lack resources to conduct audits and investigate suspected violations. Due to these limitations of public data, additional data collection strategies become more important.

**Environmental Forensics**
While RCRA and CERCLA include provisions for citizen suits, achieving an injunction against a suspected polluter in court is difficult and requires establishing defensible evidence that identifies the pollution and its source and links them to evidence of harm to health and environment.[4] Legal proof involves either physical proof of a specific pollutant, or evidence of changes in the environment that are attributable to the pollutant (Mudge 2008). To investigate such causal links, the field of environmental forensics combines methods from biochemistry, epidemiology, statistics, environmental sensing, and archival research. Environmental forensics uses an array of chemical and biological procedures to isolate the signature of a pollutant. The consequences for public health are harder to establish. The number of cases of illness is often too small to be statistically significant, or to rule out other causal or confounding factors. A final challenge is the evaluation of complex scientific evidence by a jury. Presenting forensic data for the courtroom has become an independent field of practice involving

litigation lawyers, visualization experts, and professionals from specific areas. The question of what kind of representations constitute evidence is ultimately settled by what the court and jury accept. Scientific evidence, therefore, cannot be separated from rhetorical aspects.

## Popular Epidemiology

Social action driven by affected citizens has often played a decisive role in investigating toxic pollution. Such was the case of Woburn, Massachusetts, where residents fell victim to unusually high rates of childhood leukemia and other rare diseases. Based on the noxious smell and taste of the town's drinking water, Woburn families suspected toxic pollution and confirmed the existence of a leukemia cluster, linking it to industrial facilities in the vicinity of the city's water supply. While their public authorities remained unsupportive, the residents generated a public outcry and worked alongside Harvard epidemiologists to gather statistics and drive the scientific investigation, producing impeccable data.

Based on this case, sociologist Phil Brown coined the term "popular epidemiology" to describe the engagement of citizens in investigating issues of pollution and their consequences for human health. He offers the following definition:

Popular epidemiology is defined as the process by which laypersons gather statistics and other information and also direct and marshal the knowledge and resources of experts in order to understand the epidemiology of disease. ... These data are used to explain the etiology of the condition and to provide preventive, public health, and clinical practices to deal with the condition. Popular epidemiology includes more elements than the above definition in that it emphasizes basic social structural factors, involves social movements, and challenges certain basic assumptions of traditional epidemiology. (Brown 1987, 78)

While this form of investigation is not antiscientific, it nevertheless deviates from the traditional ideal of value-free judgment delivered from a dispassionate position, which may help to prevent bias but also diminishes the impetus to address real-world problems. Without popular involvement, many health hazards would never have been discovered (Brown 1987, 87). Or as the economist Matthias Ruth puts it, "if society points at a problem, science will study the finger."[5]

## Citizen Science

Brown revisited his paper ten years later and expanded the term "popular epidemiology" into today's more familiar concept of "citizen science" (Brown 1997). "Citizen science" currently has at least two different meanings. First, the term describes investigative practices in which amateurs collect data for later scientific analysis, often under the supervision of scientists.[6] While such projects often have an educational purpose, they can serve a need under circumstances when researchers cannot gather the needed data alone (Trumbull et al. 2000). Such was the case when a container with 29,000 rubber ducks fell from a cargo ship into the North Pacific Ocean, and oceanographers Ebbesmeyer and Ingraham enlisted a global community of volunteers to report where the toys washed up. The locations of 400 toys recovered on coasts from Alaska to the British Isles were instrumental for calibrating models of oceanic currents (Ebbesmeyer et al. 2007).

A second, more assertive understanding of citizen science sees the practice not as subordinate to the work of scientists, but as an alternative model for knowledge production. "Grassroots science" aims at democratizing science while challenging the disciplinary boundaries and institutional politics of traditional scientific practice (Hansen 2005; Irwin 1995; Heiman 1997). The primary goals of grassroots science are societal rather than academic, as can be demonstrated in the practice of the Public Lab, a community organization that develops DIY tools for environmental monitoring. In the wake of the 2010 Deepwater Horizon oil spill in the Gulf of Mexico, many affected residents faced insufficient access to information about the extent of environmental damage and the effectiveness of the official disaster response. In this environment, Public Lab organized volunteer groups to collect information in the affected areas and developed open source hardware tools and methods to assess the situation. Such methods include aerial photography with kites and balloons and DIY spectroscopy using the image sensors of cell phones to detect crude oil pollution in water samples (Dosemagen et al. 2013). The work of Public Lab was motivated by the frustrations of residents who felt ignored by public agencies as the environmental disaster unfolded. It aimed to provide a collaborative space where scientists and citizens worked together.[7]

### DIY Sensing Technologies

Tinkering with sensing technologies to develop accessible alternatives to expensive, lab-grade diagnostic sensors is a common element in many similar initiatives, which often involve collaborations between technologists and social activists. The Safecast initiative developed a smartphone-based Geiger counter for mobile radiation mapping to address the shortage of radiation measurement equipment after the disaster at the Fukushima Daiichi nuclear power plant. The combination of the radiation sensor with GPS technology allowed the creation of detailed radiation maps by georeferencing measurements while investigators drove through the affected areas to avoid prolonged exposure to radiation. As a result of their approach, the team called attention to methodological flaws in official practices of radiation measurement (Bonner 2012).

Although motivations to develop novel technological solutions align with the social or environmental goals of participants, it might not always be clear which motivation dominates, considering the recent wave of innovative technological projects that use DIY drones to map dumpsites and floods, harvest Twitter to generate data for emergency responses, or build submersible sensors to inspect sewer systems. Through practices of reverse-engineering or hacking, such initiatives offer alternative approaches to scrutinizing infrastructural systems that engage directly with their materialities and black-boxed mechanisms. Since these practices of DIY sensing often take place within the environment and social contexts where their data are needed, they can be sensitive to the needs of a particular context rather than taking a neutral, seemingly objective stance.

Citizen science initiatives can augment and challenge official data infrastructures, offering alternative readings of situations and events. They call attention to the process of measurement in both formal monitoring procedures and the work of volunteers themselves. Practices of citizen science make evidence experiential by making all stages of data production explicit and tangible, from capturing traces, to encoding and validating data sets, to establishing causal connections.

Research produced by citizen science initiatives often struggles for legitimacy due to the involvement of amateurs with an interest in a particular outcome, their DIY methodologies, and their inherent data quality issues, and more generally because of the probabilistic nature of the investigated phenomena, which can render evidence uncertain and anecdotal. It should be mentioned however that epidemiology often faces similar struggles vis-à-vis other fields since its methods are limited to correlational observations and inferences that can be easily challenged.

## Conceptualizing Infrastructure Legibility

In the previous sections, I used the expression "reading infrastructure" to describe practices of data collection by regulatory agencies and their consulting firms, companies engaged in waste management, and the observations of citizens using a system.

In the following sections, I will develop this notion of infrastructure legibility as a shared framework for practices and technologies through which infrastructural systems are made legible and are used to inform infrastructure governance.

Using the example of TRI, CERCLIS, and other official databases that have played roles in infrastructure controversies, we can identify the following dimensions in which the waste system is captured:

*Structures and processes:* the geography of facilities, transportation routes, sites of discharges and disposal. Furthermore, the processes of interest that take place in and between these localities.

*Actors and consequences:* the participants in the system, including users, firms, and institutions, along with the consequences of their actions. The health effects of toxic releases, the environmental effects of greenhouse gas emissions, and the overall social and economic costs.

*Governance and the individual:* this includes documentation of court decisions and civic initiatives that have shaped environmental policy. The explicit provision that any person might bring an environmental lawsuit points to the role of the individual in the larger system—what is the scope and the consequence of individual engagement, and what motivates it?

As argued in the previous sections, the material, social, and organizational practices of technologies are intertwined across these dimensions. The databases that hold these different data points create a world (Dourish 2014). They shape the practices through which they are populated and maintained, reinforcing ideas about how things relate to each other. In a similar way, practices of legibility such as official monitoring and DIY sensing reshape the system under scrutiny and mediate its experience through their own lenses.

## About Legibility

The term "legibility" has a long history in the fields of design, urban planning, and governance. Legibility is different from visibility and invisibility. It does not depend on awareness, but on the ability to differentiate features and the capacity to detect and assemble a coherent picture from them. For a typographer, legibility refers to the ease with which a reader can distinguish the individual characters of a typeface (Lieberman 1967). Legibility can come with invisibility; the most legible typefaces are the ones that become transparent to the reader. In the spirit of modernist functionalism, typographer Beatrice Warde compared the design of a typeface to that of a wine glass—not drawing attention to itself but to its content (Warde 1955).

In the context of urbanism and governance, the term "legibility" is closely connected to the urban designer Kevin Lynch and the anthropologist and political scientist James C. Scott (Lynch 1960; Scott 1999). Both use the term in unrelated contexts. Yet how they conceptualize legibility is of interest in this book since they represent different modalities of knowledge production. The same data set can result from very different practices. It can be routinely collected by a public agency, assembled by forensic researchers, or pieced together from individual observations by a group of activists. Making a system legible can involve piecing together clues in an emergent, bottom-up process. It can also mean introducing universal standards for data collection, which continue to shape and structure the practices in the system.

**Legibility from Below**   In the mid-1950s, Lynch and visual artist Gyorgy Kepes began an investigation of *the Perceptual Form of the City* (Lynch 1955), hypothesizing that the perceived visual quality of a city depends on how well its inhabitants can recognize its spatial structure and construct an image of it: a good city is legible. In his seminal *The Image of the City,* Lynch (1960) defined "legibility" as "the ease with which its parts can be recognized and can be organized into a coherent pattern" (2). By being able to recognize and relate urban elements abstracted in his quintet of *paths, edges, districts, nodes,* and *landmarks,* urbanites construct a mental image of their environment, which supports their orientation and navigational tasks. Throughout his career, Lynch revisited the concept of legibility from different perspectives, such as the legibility of time and change in the urban environment, or the legibility of consumption and waste (Lynch 1972, 1991). In the Lynchian view, the legibility of the city structures the urban experience. "Spatial legibility is at least a common base around which groups can cohere and on which they can erect their own meanings" (Lynch and Hack 1971, 74).

Beyond transportation networks such as roads and the subway system, the role of urban infrastructure for mental representations of the city did not seem to hold specific interest for Lynch, although infrastructural elements kept emerging in sketch maps and interviews he and his collaborators conducted with residents in several cities. Furthermore, Lynch largely excluded the role of symbolic representations, narratives, and media images for making sense of urban systems, since he felt that the introduction of meaning would have complicated the investigation.

While the physical design of waste infrastructure is not the primary concern in this project, it is nevertheless interesting to consider from a Lynchian perspective how the waste system is imagined, and which elements contribute to its mental image: practices of consumption and discarding, interactions with public services, educational flyers from local governments, investigative media reports, the architecture of transfer stations, and so on. Such experiences of the waste system are explored by several scholars and artists (Nagle 2013; Royte 2005; Jackson 2011). But the sensory experience and imagination of the waste system is rarely connected to practices of monitoring and data collection.

**Legibility from Above**   In his book *Seeing Like a State*, James C. Scott conceptualizes legibility from the perspective of large-scale data collection. For Scott, legibility is a fundamental problem underlying the governance of a modernist state: knowing where things and people are. To achieve this goal, the state has to make society legible by arranging the population "in ways that [simplify] the classic state functions of taxation, conscription, and prevention of rebellion" (Scott 1999, 2).

Scott's concept of legibility is prescriptive more than descriptive, setting norms "that organize people's daily experience precisely because they are embedded in state-created institutions that structure that experience" (Scott 1999, 83). In this sense, establishing legibility is an exercise of authority similar to Foucault's notion of governmentality, exemplified by prisoners inside a panoptic prison who cannot see the guard, but are aware that the guard may be watching them at any moment (1977, 170).

Scott's production of legibility relies on standardization and simplification, which in his view motivate the institution of surnames and birth registers, universal units of measurement, and cadastral maps. Each of these institutions replaces vernacular conventions and therefore allows an administration to quantify a society and its economic output at a larger scale across space and time.

**Simplification and Superstition**   Besides his critique of how state institutions exert power through data collection, Scott's perspective of legibility as top-down data collection is also relevant for understanding the technicalities of waste monitoring. Simplification helps reduce and focus information, to "summarize precisely those aspects of a complex world that are of immediate interest to the mapmaker and to ignore the rest" (Scott 1999, 87). When data are used to summarize and compare across space and time, simplification is more important than completeness: Scott observes: "A city map that aspired to represent every traffic light, every pothole, every building, and every bush and tree in every park would threaten to become as large and as complex as the city that it depicted. And it certainly would defeat the purpose of mapping, which is to abstract and summarize" (Scott 1999, 87).

However, such maps now exist, and the role of simplification has changed in contemporary data analysis. For all practical purposes, spatial data can become as complex as the territory they map when the data comprise administrative data sets, historical archives, the digital exhaust of information infrastructures, real-time sensor feeds, and information submitted by or collected from individuals. The data these maps contain are often impure because they are collected for unrelated reasons and under various circumstances, subject to all kinds of biases. For a traditional statistical analysis that relies on randomized samples collected under controlled conditions, they would be of limited use.

Contemporary methods of analysis embrace the messiness and heterogeneity of data aggregations that are considered biased and incomplete by definition. In Bayesian statistics, statements of true and false become fuzzy, replaced with probability distributions inferred from the data at hand. According to Viktor Mayer-Schönberger and Kenneth Cukier (2013), "more trumps better," meaning that larger data sets can offer better capacity to control for biases and evaluate a large number of hypotheses and model specifications. Exploratory statistical methods are designed for loosely structured data aggregations that resemble the second type of map Scott describes, which provides a contrast to traditional administrative simplifications:

The second map consists of tracings, as in a time-lapse photograph, graph, of all the unplanned movements—pushing a baby carriage, window shopping, strolling, going to see a friend, playing hopscotch on the sidewalk, walking the dog, watching the passing scene, taking shortcuts between work and home, and so on. This second map, far more complex than the first, reveals very different patterns of circulation. The older the neighborhood, the more likely that the second map will have nearly superseded the first, in roughly the same way that planned, suburban Levittowns have, after fifty years, become thoroughly different settings from what their designers envisioned. (Scott 1999, 347)

The promises of Big Data are often overstated, and despite the capacity to control for biases in large data sets, the principle of "garbage in—garbage out" still holds. Biases are not always the result of a statistician's prior beliefs and assumptions. Often they are artifacts produced by the statistical models themselves: extrapolations of patterns caught by machine-learning algorithms, which might be causally meaningful or just products of chance. Constructivist philosopher Alexander Riegler described this sensitivity to any pattern, meaningful or meaningless, as "superstition in the machine" (2007), which can, for example, make Google Adwords appear racist when its algorithms extrapolate from hidden patterns in collective search behavior (Sweeney 2013).

Opportunistic data sources offer additional avenues to read the waste system beyond traditional data sets. Location-based media services are used as proxies for human behavior, and the distribution of people in public spaces can be used to estimate consumption and waste production. Utility companies embrace wireless sensors in public waste bins and RFID chips in private trash cans to measure waste production in order to promote a change to performance-driven contracts. Instead of fixed collection routes and schedules, municipal contracts would specify data-driven standards that require servicing only those cans that need to be emptied.[8] Besides the advertised cost savings for cities, such a shift would also move sensors and data-driven standards into the center of infrastructure governance and have implications for how contracts are negotiated. Whoever has control over sensors and their data will have an advantage in such negotiations. Infrastructure monitoring is no longer limited to Scottian power mechanisms of simplification and standardization. But these more agile and responsive modes of reading the city through real-time data introduce their own politics and power differentials.

**Spatial Information as Text and Trace**   The term "legibility" implies an act of reading, which at first seems to imply the metaphor of the city as a text. Reading infrastructure in this sense relates to the previously described practices of data analytics and urban informatics, which operate in the symbolic space of data sets from different sources. Legibility in this sense means reflecting on the languages, classification systems, and symbolic operations. The language used to describe the waste system has to be universal enough to allow aggregation and computation yet specific enough to frame the subject matter for different perspectives of the actors involved. This symbolic legibility also concerns how and in what forms such information is made accessible to the public.

But the act of reading is not necessarily limited to symbolic systems. Reading also describes the process of collecting physical traces in the environment: Latin *legere* and the German *lesen* also have the meaning of *gathering, picking up*. The trace is not a symbol, but an imprint, an index pointing to the event that created it (Peirce Edition Project 1998). Studying traces invokes the realm of forensics, meaning that if one looks closely enough, there are no two things in the world that are identical (Kirschenbaum 2008). But traces also need to be made visible and they require interpretation. Traces are not "out there" waiting to be found, but are the result of a method of observation: before one can observe tree rings, one has to cut a tree.

The example of GPS coordinates illustrates the distinction between these two aspects of reading. On the one hand, coordinates are a set of alphanumeric characters based on the encoding convention of the World Geodetic System of 1984 (abbreviated WGS 84). Geographic coordinates allow mathematical operations such as calculating average distances, clustering into distinct regions, smoothing and simplifying trajectories. On the other hand, they bear, like all media, characteristics of material traces (Dourish and Mazmanian 2013). Their acquisition depends on physical processes, technical mechanisms, and environmental conditions such as an unobstructed line-of-sight connection to at least three NAVSTAR satellites, their relative locations to allow triangulation of sufficient accuracy, and last but not least, a charged battery on the receiver.

Beyond their capacity for quantitative analysis, GPS traces gain implicit meaning relative to the objects, places, and facilities in proximity. A location sent by a tracked soda bottle from within the boundary of a landfill might indicate the bottle's disposal, although other interpretations are always possible. This type of analysis is a qualitative judgment, an act of interpretation that can be supported by computational classification methods since inferring meaning from locations can be easier and more reliable than from other methods. For this reason, spatial metadata are increasingly used to provide additional context in the analysis of traditionally nonspatial data.

Using tracking and participatory sensing to read waste systems therefore calls for a mixed-methods approach that is not limited to the values in the data set, but also considers the circumstances of data collection and contextualizes the data with other sources of information.

## Elements of Infrastructure Legibility

Under what circumstances can local actors benefit from DIY practices of legibility? To what extent can such practices enrich, contrast, or challenge official representations of the waste system to make them more inclusive? Scott, who referenced Lynch in a single footnote, suspects that legibility from the perspective of the inhabitant and legibility from the standpoint of the administrator are negatively correlated (Scott 1999, 385). The schemata of large-scale data collection are too crude for understanding local conditions in their complexity, while the idiosyncrasies of local knowledge resist large-scale aggregation and comparison.

In an environment where administrative schemata and policies do not represent local needs adequately and institutions are inaccessible, resistance by obfuscation or a refusal to share information might be the only options to preserve local knowledge and practices. This is relevant in environments where informal waste management is criminalized or subject to repressive formalization policies, an issue discussed further in part II of this book. One might argue that the informal waste sector still exists and is economically relevant not despite its lack of legibility, but precisely because of it. Similarly, data collection through DIY waste forensics might be ineffective when the institutions and the legal system are inaccessible to citizens who want to address cases of pollution.

Conversely, in an environment where public opinion and civic action are potential drivers of environmental policy, local initiatives can benefit from collecting data through coordinated action, combining and sharing through open standards. Environmental groups have engaged in large-scale data collection and successfully used their findings in federal courts to establish or defend environmental policies or to instigate regulatory action. Administrative taxonomies that are necessarily rigid in some aspects can be malleable in others, accommodating and sensitive to local knowledge and needs. Part III of this book discusses examples where official classifications of public services were adapted based on citizen input to reflect better how the public perceives urban problems.

Collecting and opening data to the public is no panacea. Just as states often shroud themselves in secrecy and corporations withhold proprietary information, tactics of obfuscation, refusal, and invisibility can be important instruments for bottom-up collective action. But in each case, the literacy of systems, technologies, and representations of their interconnections are of critical importance.

In the following section, I review the six elements of infrastructure legibility, which I introduced earlier in an ad-hoc taxonomy.

## 1.  Structure
Reading the spatial fabric of a system often requires a map, yet many waste facilities are missing from popular online mapping services.[9] The geography of the waste system concerns—among other aspects—the locations of facilities and their interconnections: landfills, waste-to-energy plants,

transfer stations, public waste bins, or transportation routes. Federal databases catalog facility locations (U.S. EPA 2009) but do not include information about their interconnections and roles in the larger system. Such
information can sometimes be gleaned from municipal service contracts,
while information about the structures of industrial waste management is
harder to come by because it is scattered across different repositories, each
of which reflects a partial aspect of the system.

## 2.  Process

Waste systems are not defined through their spatial structure, but more
importantly by their processes and flows. Garbage collection, processing,
and disposal are primarily temporal phenomena. In this context, media
scholar Ethan Zuckerman distinguishes between two kinds of maps: infrastructure and flow. Infrastructure maps show the space of what is possible, while flow maps show what actually happens at a given point in time
(Zuckerman 2013).

   Real-time flow maps of public service provision are typically reserved
for control rooms in public works and sanitation departments. Yet such
information can change the public perception of a system: the introduction of real-time countdown timers in subway stations has reduced the perceived waiting time for trains (Chow, Block-Schachter, and Hickey 2014).
To address public concerns during snowstorm emergencies, New York
introduced real-time maps of snow removal.[10] Real-time feedback can profoundly change activity in the system. Real-time traffic information may
help to avoid traffic jams but can also create them.

   Yet many processes remain practically unobservable in the waste system.
International flows of electronic and hazardous waste are poorly understood, and as a result, waste legislation and international agreements may
end up promoting or banning the wrong practices (Lepawsky 2014, 11).

## 3.  Actors

Every system is maintained by people, yet most diagrammatic representations of infrastructure conspicuously leave out its managers, workers, and
users. Nevertheless, most infrastructures not only offer clues about their
users' presence, but they also provide a medium through which people
communicate and interact.

   Garbage trucks bear slogans promoting recycling and resource conservation. Public waste bins are often found covered with posters and flyers. An
overflowing waste bin indicates human presence and activity; the discarded
objects are records that provide further clues about these activities. Garbage

collection and municipal services are often the most visible and immediate ways through which a local administration can present itself.

Identifying and shaming responsible actors is a widespread strategy for advocates and activists to expose user behavior and encourage change. The creators of the "bincam" project encourage users to share photos of what they throw away in order to engage their friends in a competition over who is most environmentally responsible (Comber et al. 2013). Disclosing how users consume public services raises privacy concerns, however. A design strategy that addresses these concerns makes systems "translucent" rather than "transparent"—similar to how a frosted glass door conveys activity in a room without revealing its occupants (Erickson and Kellogg 2000).

## 4. Consequences

The social, environmental, or economic consequences of a particular waste management process are often difficult to establish, and grievances can rarely be linked to a single cause or event. Answering questions of causality and impact is possible only if detailed data about all processes and aspects are available. Their analysis requires converting all influencing factors into a shared unit of comparison: cost, greenhouse gas emissions, footprints of land, energy, or water consumption.

Life-cycle assessment (LCA) offers a comprehensive approach to estimating the environmental impact of a product, facility, or process. LCA involves hundreds of parameters and models for comparing different scenarios such as landfilling, recycling, or incineration, yet their outcomes can be conveniently visualized through their common denominator, for example an energy footprint. The simplicity of the results hides the complexity of the evaluation process and lets us forget that many parameters are based on simplified assumptions.

Identifying the responsibilities of consumers, governments, and manufacturers is an everyday aspect of environmental struggles. Representations are not neutral in these controversies; they can be used to shift the blame. Recycling rate metrics are sometimes criticized for shifting responsibility to the individual and local governments while exonerating producers. As anthropologist Max Liboiron notes, "Recycling is rarely represented as an industrial process, or as a form of waste management. Instead, its primary meaning comes from its status as a kind of environmental activism" (Liboiron 2009).

## 5. Governance

"Sunlight is said to be the best of disinfectants; electric light the most efficient policeman," observed U.S. Supreme Court Justice Louis D. Brandeis

(1914). "Transparency" is the battle cry of making governance legible. While the U.S. Freedom of Information Act (FOIA) and Community Right-to-Know provisions have played an instrumental role in social initiatives such as the environmental justice movement, the focus of open data initiatives is on the details of everyday governance processes rather than large conflicts.

Through the help of digital platforms, the open data movement aims to bring citizens and local governments together to solve urban problems in a more responsive and participatory manner. The promise is that government data in machine-readable formats will not only enable new services and applications that benefit the public, but will also allow evaluation of policies based on their outcomes rather than on adherence to rigidly defined process (Goldsmith and Crawford 2014). Ideally, collection methods for administrative data are open to scrutiny. In the United States, however, a significant portion of federal data on MSW is collected by private consultants using undisclosed and proprietary methods (MacBride 2012, 18).

Transparency alone, however, cannot ensure good governance and prevent corruption. As noted by Aaron Swartz, the late data activist, simply putting databases online will not lead automatically to meaningful democratic dialogue. Ironically it can also be a way to hide information by drowning relevant issues in a torrent of unrelated data (Swartz 2010). Furthermore, since transparency can highlight failure over success, it can be easily misused as a political instrument (Ben-Shahar and Schneider 2014; Lessig 2009).

I find the term "accountability" more useful for making governance legible, since it highlights relationships and dependencies between actors. "Giving account" is an act of communication and interpretation rather than a passive means of default. Part III of this book will discuss the relationship between governance and accountability, and its legibility in data representations and interfaces.

## 6.  The Self

Open data portals and citizen reporting apps not only make the city and its governance legible, but they also act like mirrors that reflect their own roles in infrastructure management. Seeing oneself reflected as a contributor in official representations of the city can be a major motivator for civic engagement.

In many states, Adopt-a-Highway programs engage volunteers to clean up roadside litter, rewarding involvement through a sign that bears the donor's name. In a similar arrangement, New York City residents can adopt a wastebasket and take responsibility for emptying it on a regular basis. SeeClickFix, a civic digital communications platform for monitoring infrastructure, dedicates ample space to participants by including activity indicators, self-written user profiles, and other design features that create symbolic reward mechanisms found in volunteer-driven communities such as Wikipedia.

On the one hand, designing these platforms requires an understanding of motivation. How do volunteers experience the system? How do they view themselves in the system? Do they hope to learn new skills, find new friends, or share a cause? How are they rewarded for contributing? On the other hand, the design influences the platform's use. Inclusions and omissions in the feedback system influence user actions. In sociologist Steven Woolgar's term, design "configures" the user (Woolgar 1991).

## Conclusion

These six aspects of infrastructure legibility do not form an exhaustive list. They do not consider time and sequence, which could be a cross-cutting seventh element of infrastructure legibility. Temporal legibility includes archival traces such as server logs and databases that record interactions on a system. Most open data platforms offer historical data, yet it should not be taken for granted that records are always preserved. Public administrations frequently do not have the funds for data warehousing, and companies often keep only data deemed valuable to them.

The following chapter describes an attempt to read the structure and processes of a system for which no comprehensive data are available. By tracing the path of individual waste items across system boundaries, the study aimed to reconstruct the geographies and the commercial and industrial topologies of the waste stream. This experiment provides nothing more than a snapshot. For all practical purposes, no sample is sufficient to infer a reliable picture of a municipal solid waste system, which in itself constitutes only a tiny fraction of the volumes of industrial waste. It nevertheless offers a glimpse from the outside, and a view that is not limited by organizational boundaries but instead by the modes of digital representation and by the resilience and longevity of the sensor.

## 2  Reading Structure in Waste

We cannot turn our entire planet's crust into obsolete objects. We need to locate valuable objects that are dead, and fold them back into the product stream. In order to do this, we need to know where they are, and what happened to them. We need to document the life cycles of objects. We need to know where to take them when they are defunct. In practice, this is going to mean tagging and historicizing everything. Once we tag many things, we will find that there is no good place to stop tagging.

—Bruce Sterling's keynote presentation at the 2004 SIGGRAPH conference (Sterling 2004)

The waste system is not monolithic. It consists of many components that communicate and interact with each other while remaining in separate realms. Among its components, we find transfer stations, material recovery facilities, specialized recyclers, landfills, compost facilities, waste-to-energy plants, and more. Owned and operated by cities, private contractors, and increasingly, multinational corporations, the waste system has many seams that separate the physical and informational domains of companies, administrative areas, and regulatory authorities. All of these actors have built their own information infrastructures for keeping track of processes and material flows. However, little of this information is accessible to the public or exchanged across the system, and what is shared cannot always be integrated into a coherent image. Based on the available information, it is hard or impossible to track objects as they move across system boundaries.

Conducted from 2009 to 2010, the Trash Track study started with a number of simple questions. Where does waste go? How far does it travel, and to what extent does its transportation diminish the benefits of recycling? What would be the effect if everyone could see what happens to his or her own waste? Accomplished with the help of local volunteers, the project was an experiment to map the trajectories of waste and recyclables through the global waste system at the scale of the individual item. Before discarding items, participants attached electronic sensors that would report geographic locations at regular intervals during the items' journeys through the waste stream. The project addresses the question of how the subsystems for waste hauling, material separation, recycling, and disposal are connected—not only in their design, but also through their interactions. By doing so, it explores a central aspect of infrastructure legibility: the capacity to read the structure and processes at the whole-system scale.[1]

## Seattle's Waste System

Seattle is perceived as an environmentally responsible city, although it has had its share of waste-related controversies regarding the siting of a proposed landfill for the city's garbage in the rural eastern parts of Washington State. The city contracts the private waste management companies Waste Management Inc. and Republic Services (formerly Allied Waste) to handle waste disposal, recycling, and composting in different service areas. Like many U.S. cities, Seattle uses a single-stream system for collecting recyclables, allowing metal, glass, paper, and plastic (except for single-use carryout bags that retailers are banned from providing) to be mixed in the same container and collected by a single-compartment truck. From the nine million tons of solid waste generated in Seattle in 2009, about 53 percent was diverted for recycling (Washington State Department of Ecology 2010).

Seattle's main recycling contractor operates a material recovery facility (MRF) one mile south of the downtown. For most objects, this is only the first stop. The city is well connected to global transportation networks, including rails and roads, with a large seaport that offers direct shipping routes across the Pacific Ocean. According to the city's contracts with different waste management providers, most plastics and between 70 and 100 percent of its paper and cardboard waste collected from the curbside are exported to Asia, primarily China (Seattle Public Utilities 2003). Some categories of waste, including electronic waste (e-waste) and household hazardous waste (HHW) items such as computers, compact fluorescent light (CFL) bulbs, or TVs, are not covered by curbside collection. They can be brought to retailers and other businesses, recycling centers, or special collection events. CFLs can also be mailed back to recycling centers in pre-paid envelopes. The remainder of the municipal solid waste is hauled by train about 400 kilometers to Columbia Ridge, an arid, sparsely populated region close to the Columbia River near Arlington, Oregon. According to its operator, this landfill collects two million tons of waste annually, and the methane it captures powers 12,500 homes (Waste Management Northwest 2015). Close by, there is also a compost facility for Seattle's curbside-collected food and yard waste.

## The Trash Track Experiment

Tracking the paths of waste with self-reporting location sensors seems like an almost trivial proposition: attach sensors to a garbage object, and start recording. The resulting traces promise a window into the minute details of waste systems from the outside and support or contradict existing assumptions, claims, or institutional data sets. Waste tracking therefore allows the scrutiny of infrastructural systems and their representations from a different angle. Its critical potential is not limited to unpacking existing institutional data sources and arguments, but involves building technologies and generating information that facilitates critique. Such an approach could be seen as an example of what has been described as *critical making* or a constructive approach to critique (Ratto, Boler, and Deibert 2014; Wylie et al. 2014).

As I will show in this chapter, the method turned out to be more complicated than suggested by the simple premise. It had several limitations, most of which were related to the scope and the unwieldy material reality of the waste system, which is not a hospitable place for experiments with sensitive electronic devices. Considering the substantial physical strain on the sensors over an extended period, the resulting window into the interior of the waste system was tiny, and the interpretation of global positioning system (GPS) traces was not possible without additional data sources. Identifying facilities based on geographic coordinates required manual analysis and access to aerial images and facility registries. Observed waste flows had to be contextualized with the information found in trade databases or waste management contracts. Waste tracking is an expensive and labor-intensive process, and therefore, it is hardly possible without support from outside actors.

With the help of volunteers recruited through local media, our team located appropriate garbage items in different parts of the city and outfitted them with sensors. From the pool of 105 households, ninety households and six school classes were selected based on their spatial distribution, involving about five hundred individuals. Each household was asked to prepare fifteen to twenty items according to a prioritized list that ensured

**Figure 2.1**
Volunteer with donated objects and sensors before tagging. Photo: Christophe
Chung, courtesy MIT Senseable City Lab, 2009.

a broad range of objects, including magazines, textiles, glass, corrugated
cardboard, plastic containers, batteries, cell phones, electronic appliances,
toys, tires, and furniture.

At the beginning of the project, the Trash Track team only had a very
general idea about the role of the public in this project. There were certainly practical advantages of working with volunteers—for example, as a
mechanism for getting access to a broad range of waste items and ensur-
ing the even distribution of the sampled items throughout the urban environment. For this purpose, the environmental scientists Malima I. Wolf
and Avid Boustani compiled a list of desirable waste items to mimic the
typical composition of household waste and recyclable materials. However,
the community's enthusiastic response, ideas, and contributions quickly
brought the participatory aspect into the foreground of the project. The
volunteers had different motivations. Some were excited about the possibility of making waste transportation more efficient and sustainable. Others
saw the experiment as a method to hold the city accountable for its promises and actions. The project held personal significance for some people

who wanted to understand the consequences of their consumption patterns. Some participants even brought objects that they wanted to dispose of, but were still attached to. Trash Track gave them a way to say farewell by following the objects on their final journeys.

Our team members assisted the volunteers with attaching the sensors and recording information about the objects. After the sensors had been deployed, the volunteers were asked to dispose of the items as they normally would. This not only included regular curbside collection, but also any process the volunteers deemed appropriate, including bringing items back to retailers, participating in public collection events, or using public waste bins. Because garbage items come in a variety of materials, shapes, and sizes, attaching sensors required skill and improvisation, contradicting the notion that digital technology produces data as an effortless byproduct. The biggest problem was protecting the sensitive electronics of the sensors from the inhospitable environment of the waste stream. Many factors had to be considered, such as the crushing physical forces inside the waste stream, moisture, and signal blockage by buildings, the metal body of a collection truck, or waste materials themselves. Notably, the metal housings of appliances such as DVD players or PCs were challenging in this respect. How to prevent the tracker from detaching also had to be carefully considered. The early phase of the experiment was therefore dedicated to finding a suitable technique for attaching sensors to a broad variety of objects and materials as well as protecting the sensor without blocking the signal, while making sure both sensor and object became practically inseparable.

It quickly became clear that many solutions fulfilling all of these requirements were not practicable in the field, as they were too complex in their execution or required too much time. The approach chosen was to use two-component, expanding epoxy foam to attach the sensor to the object right on site and to encapsulate each sensor in a one- to two-centimeter shell of the material, which in preparation required only a few minutes to expand and become rigid. The lightweight material was durable and transparent to radio signals, protecting the tag from being crushed and insulating it against water. The foam, originally intended for repairing boat hulls and surfboards, required a swift and skillful application, and a different application strategy was needed for each object. To preserve the original appearance of discarded items, items smaller than the tracking devices were excluded. Organic waste was also excluded to prevent sensors and epoxy from contaminating the compost. These preliminary experiments were crucial to the success of the project. While in preliminary tests almost 80 percent of sensors were quickly destroyed and never reported a useful trace, it was possible to lower this number to about 20 percent as the experiment progressed.

## Technologies for Waste Tracking

The speculative idea of smart trash, garbage that self-reports its location and material composition, has been explored by environmental scientist Valerie Thomas and her collaborators, adapting localization and identification methods that were already used in supply chains (Lee and Thomas 2004; Saar and Thomas 2002). In the United States, the Basel Action Network (BAN) investigated illegal exports of e-waste to China, all of which gained national attention in an episode of the popular CBS TV program *60 Minutes*, while Greenpeace used GPS receivers to track the illegal export of TV monitors from the UK to Nigeria (CBS News 2008, Greenpeace International 2008). Since the completion of the Trash Track study, BAN has used self-reporting GPS sensors to monitor the actions of e-recyclers across the United States and collected evidence of otherwise undocumented

**Figure 2.2**
Tests of different materials for the protective enclosure of sensors, from left to right: epoxy foam (eventually used in the experiment), rubber, and epoxy resin. Photo: Jennifer Dunham, courtesy MIT Senseable City Lab, 2009.

exports of monitors and printer cartridges to Asia (Basel Action Network 2016).

Tracking more than a few objects in a singular effort raises many technical and methodological challenges. In the subsequent section, I discuss in depth the different technological options for tracking waste. For the Trash Track experiment, the technology of choice had to record the geographical routes of a sufficiently sized sample of household garbage within the waste stream, capturing as much information as possible about each item, independent of location and mode of transportation.

Tracking technologies can be divided into two categories: methods for identification and methods for localization. At an industrial scale, most automatic tracking is done using the inexpensive radio-frequency identification (RFID) chips that are found in price tags and shipping labels as well as in access cards and subway passes. These tags can be detected by an interrogating device at close range—from less than a centimeter up to a few meters. Since these chips typically do not carry their own power supply, they can transmit their stored data, a short sequence of bytes, only when they enter the electromagnetic field of a detection device. RFID is therefore a method for identification that allows localization only implicitly through the known location of the reading apparatus. The postal service or global shipping companies such as FedEx command and operate an extensive network of reading devices built into facilities, vehicles, and handheld devices, therefore allowing the accurate localization of items. Most waste providers do not use such technology, ruling out this relatively straightforward and inexpensive option. A second disadvantage for the purposes of waste tracking is that the detection of RFID tags can be easily avoided.

Most methods for localization require active sensors equipped with a power source and a transponder capable of communicating across longer distances. In a reversal of the method just described—readers at known locations scanning their surroundings for detectable items—the active, mobile sensor scans its environment for known reference objects. These objects can be satellites broadcasting radio signals—like those of the U.S. GPS, the Chinese BeiDou,[2] the Russian GLONASS,[3] or the European EGNOS/Galileo system.[4] The location is calculated by the receiver based on the time differences of signals from at least three satellites in the sky above the receiver. Depending on signal strength and receiver sensitivity, satellite-based localization requires an unobstructed view of the sky.

To improve the quality of localization, modern sensors using differential GPS (DGPS) or assisted GPS (AGPS) architectures do not exclusively rely on satellite signals, but also take the known locations of other reference objects into account: mobile network base stations, Wi-Fi hotspots, Bluetooth access points, or other terrestrial radio signals. This can make localization faster and more robust and work in situations when the line of sight to the satellites is obstructed, for example, indoors.

Combining global localization and close-range identification, several hybrid approaches exist. Devices used for tracking wildlife usually combine a GPS sensor for location acquisition, a wireless communication module for reporting it back, and a radio beacon that allows investigators to find and retrieve the tracker nearby. Smartphones are not only capable of sensing their own location using a variety of satellite and terrestrial radio signals, but also detect other objects in their own proximity, such as low-power Bluetooth devices or passive RFID tags at close distance. The localization layer in smartphone operating systems constantly scans the environment for identifiable objects that are then reported to Google or Apple as reference points for improving their localization service—a practice that has raised privacy concerns, as it catalogues locations of both private devices and users. Hybrid localization products such as Tile[5] use small tags with embedded low-power Bluetooth beacons that can be localized not only by one's own smartphone, but also through all other smartphones that have subscribed to the service. The nirvana of networked location-sensing technologies, described as "smart dust" (Warneke et al. 2001), involves tiny autonomous sensors that do not require batteries, as they harvest their power from the physical movement of the tracked object or the ambient electromagnetic radiation of cell phone networks. Such technologies for unlimited surveillance are still in the experimental stage and offer, perhaps for the better, no realistic option for waste tracking.

Given the almost limitless range of possible waste destinations, ranging from authorized facilities to illicit dumpsites, the required technology for the experiment had to be capable of autonomously reporting its location from remote locations outside densely populated areas. Although the GPS facilitates location *sensing*, it does not solve the problem of *tracking*—reporting the acquired location back for data collection. The wireless phone system provided by various carrier technologies offers theoretically global reach, but it is limited to areas where the wireless communication infrastructure is installed, excluding oceans and many remote locations. True global coverage is available only for cellular satellite communication, which is prohibitively expensive.

In this context, it is worth mentioning that the Trash Track team expected no realistic chance of recovering sensors after the device stopped responding from within the waste stream. In most cases, the sensor and its battery would be buried in a landfill, incinerated, or otherwise destroyed. In terms of its environmental impact, the number of 2,500 sensors is negligible in relation to the volumes of garbage produced every day. But at the same time, this amount is hardly sufficient to generate robust quantitative data about waste movement, and a more representative number of trackers could raise environmental concerns (Wäger et al. 2005).

### Conducting the Trash Track Experiment

Evaluating the different options, the team decided that sensors that use the mobile wireless network for both localization and reporting offered the best tradeoff between location independence and cost. Two types of active location sensors were selected, both of which had roughly the size and shape of a matchbox. A set of 500 sensors developed by Lewis Girod at the MIT Senseable City Lab used cell tower triangulation to calculate location and were deployed in New York and Seattle. Each sensor returned text messages containing the IDs of all base stations detected in the vicinity, allowing the locations to be calculated back at MIT (Boustani et al. 2011). The team also used a larger set of 2,500 commercial asset trackers from Qualcomm that offered the advantage of higher accuracy by using AGPS technology for localization.

Battery life was a big technical constraint. Achieving an operation time of three to six months was possible only by minimizing reporting intervals to four to six hours. Because searching for base stations and satellites requires considerable energy, various optimization techniques were evaluated, such as attempting a connection only if the device had recently moved, or storing data on the sensor until an optimal connection became available. Because each optimization technique showed disadvantages such as less reliable reporting, we decided to use regular reporting intervals, which had the advantage of a periodic heartbeat useful for diagnostic purposes.

To find the best compromise between battery life and reporting frequency, the sensors were divided into three groups, using three-, four-, and six-hour reporting intervals. As expected, the six-hour sensors had the longest average lifetime, although the differences were not large. While the sensors were theoretically able to send back locations from other countries, this possibility was limited by the diversity of global telephony standards. Battery life also made it unlikely that trackers would reemerge on other continents after extended periods of time in the oceans or in harbor facilities.

## Data Cleaning: Null Islands and Location Artifacts

The tag deployment finished in early September 2009. All sensors had stopped reporting by early January of the following year. Before analysis, the data set had to be cleared of erroneous reports, such as the false location artifacts that resulted from poor cellular network coverage. Despite the advantages of AGPS, this step was crucial, since the quality of location data depends on the presence of reference objects and varies widely. Localization can be highly accurate in dense inner cities, but not available at all in remote, rural areas. The distribution of reference objects such as cell towers can also introduce obscure error patterns and location artifacts that are difficult to spot.

Consequently, some of the reported data appeared mysterious at first, particularly since location reports did not come with a reliable estimation of accuracy. A large number of reports came from an identical location on Vashon Island in Puget Sound near Seattle. Initially, it was not clear how a large amount of waste could end up in a residential neighborhood that had no waste facilities. On closer inspection, it turned out that each report specified the location of a single cellular base station that happened to be the only tower the trackers could access while they were, most likely, in transit on a barge. As the data set of location reports grew, such location artifacts became easier to identify because they tended to appear in identical locations. Besides the site on Vashon Island, other location artifacts with identical coordinates began to emerge, and it quickly turned out that not all of them could be easily dismissed. The location artifact on Vashon was misleading, as it indicated a location on land, when the report was likely sent from a passing vessel. However, since the received location indicated the only physical reference point accessible to the sensor, it still carried some meaningful information, and the decision whether to include or discard such a questionable report was not always clear cut.

Such situations are familiar to spatial analysts. In the simplest case of a geocoding error, the geographic coordinate defaults to the origin at the location 0.0,0.0—a place off the coast of Guinea in the Atlantic Ocean. To catch such faulty locations, geographers use an imaginary landmass called Null Island, defined at this location with an area of 1x1 meters in a geographic information system (GIS). As systems for geolocalization become increasingly hybrid and involve multiple coordinate systems and methods with different default values and error conditions, geocoding errors can default to other unpredictable locations where no null islands can catch them—such as a suburban family home in Atlanta from which lost smartphones seemed to report their location, or a farm in Kansas close to the geographical center of the United States, which was used by a mapping service as a default location (Hill 2016a,b).

If a single sensor produced at least two reports, it was considered a valid trace. Items that failed to produce any traces were excluded from the data set. From the total number of 1,971 sensors used in the final deployment (two earlier deployments used smaller numbers of 500 sensors each), 1,279 reported traces longer than 250 meters. Shorter traces were excluded from data analysis on the assumption that the sensor may have been destroyed in the collection truck's compactor or its transmission signal was blocked.

In a second step, traces that did not enter the waste removal system were eliminated. These reports could have happened because volunteers forgot to discard the object or removed the sensor out of curiosity. Such cases were identified by examining the traces to determine if an object stayed at the volunteer's home.

Because sensors would eventually fail, traces were expected to reflect shorter distances than the actual trajectory of the item. Due to exposure to all kinds of mechanical forces, liquids, and layers of material impenetrable by radio signals, the waste stream is a hostile environment for electronics. The length of the reported trace rarely represents an item's full journey. A second factor contributing to the underestimation of distances was the low temporal resolution of traces, which yielded distances closer to Euclidean as-the-crow-flies intervals between stops rather than the actual road distance traveled. Reported distances were therefore considered to be minimum values. Rather than eliminating the longest traces as outliers, we treated them as fortunate cases where the sensor held out especially long. Among the different materials tracked in the experiment, electronic waste and scrap paper had the lowest failure rates, due to less hostile physical conditions in their respective recycling streams, which are less commingled with other materials and protected from water.

### Initial Analysis: Traces in S, M, L, XL

A first glance at the mapped-out traces revealed the most salient patterns and characteristics of the waste removal process. Most traces ended within a 500-kilometer radius around Seattle, with the landfills in Arlington being a frequent destination. A remarkably pronounced cluster of reports indicated the location of the MRF close to the seaport.

While the reported traces had, on average, a length of 114 kilometers, the longest trace, created by a printer cartridge, was over 6,000 kilometers

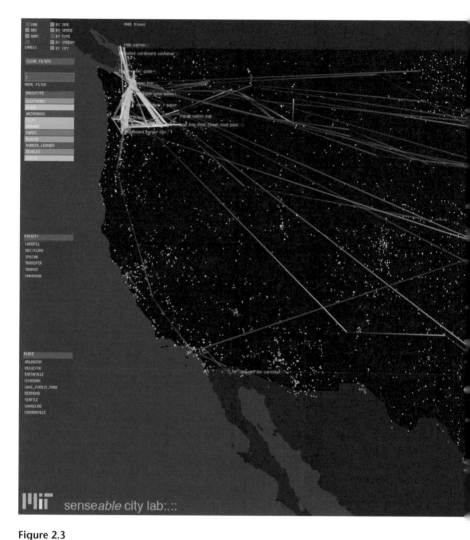

**Figure 2.3**
The collected traces overlaid with the locations of waste-processing facilities from the
EPA FRS database. Landfills are drawn in yellow, recycling facilities in blue. Visualization by the author, courtesy MIT Senseable City Lab, 2010.

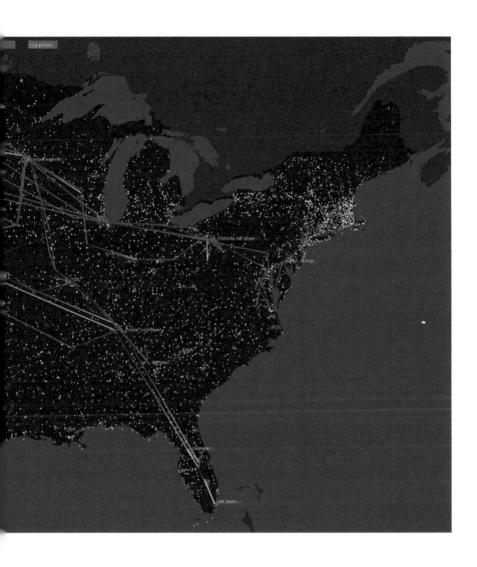

long. Reports were received from Mexico and Canada, as well as various ports on the West Coast and off the coasts of Texas and Florida. Electronic and household hazardous waste generally produced the longest traces, both in terms of distance and duration, while glass and metal items reported the shortest traces. It is important to keep in mind that these results correspond to the distance an object travels more or less unchanged, before it is processed and the sensor subsequently destroyed.

Plotting the distances on a logarithmic scale revealed three distinct clusters. The first and largest cluster included short traces with a length between 10 and 50 kilometers, mostly comprised by curbside recyclables. A second, smaller cluster emerged at a distance of approximately 400 kilometers, which corresponds to the distance to the city's main landfill. The third, smallest cluster included traces longer than 1,500 kilometers, all of them belonging to the electronic and hazardous waste categories.

While electronic and household hazardous waste items reported the longest traces, the waste categories with the highest variation in distance were all nonrecyclable wastes: the mixed waste and mixed plastic categories.

### External Data Sets for Contextualizing Results

Further analysis beyond basic quantitative summaries requires additional data sources and includes an interpretive and qualitative aspect. Since a report consisted only of a geographic coordinate and a time stamp, additional information was necessary to infer facilities and means of transport.

First, it is crucial to know whether the locations reported by the devices coincide with waste facilities. The U.S. Environmental Protection Agency (EPA) maintains a publicly accessible database of all facilities that are subject to environmental regulations, including active or closed landfills, cleanup sites, recycling facilities, and all businesses that generate, process, or ship hazardous waste (U.S. EPA 2009). From the 2.5 million records of firms monitored by the EPA, we extracted a subset of facilities involved in waste management, such as transfer stations, recycling centers, and landfills, and developed an algorithm to match the recorded locations to the locations of facilities.

The results of this automatic matching process, however, initially were not satisfactory. One complicating factor was that many facilities service multiple waste streams and the facility database sometimes listed multiple facilities and companies at the same geographic location. A second factor was that the recorded location signals contained considerable noise,

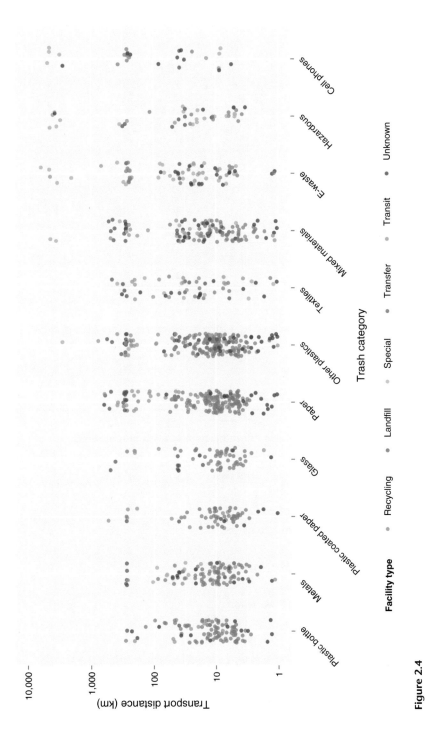

**Figure 2.4**
Logarithmic scatter plot of recorded transportation distance separated by waste category. On the vertical, logarithmic axis three distinct clusters associated with different waste streams can be identified at the distances of 10km, ~600km, and >1,500km.

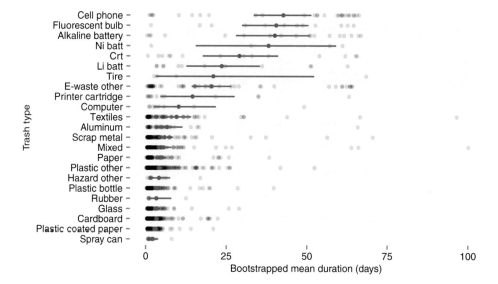

**Figure 2.5**
Mean durations of removal by waste type, with confidence intervals from bootstrapping[6] (red).

**Table 2.1**
Used public and private data sources

| | |
|---|---|
| Collected | |
| | Traces of household waste and recyclable items plus metadata |
| Federal | |
| | EPA Facility Registry System data |
| State | |
| | Disposal data and tipping fees |
| | Facility input/output tonnages |
| Municipal | |
| | Solid waste collection contracts |
| | Processing contracts |
| | Long haul disposal contracts |
| | Solid waste and recycling reports |
| Private | |
| | RecycleNet Spotmarket Scrap prices 2010. Source: http://recycle.net/spotmarket |

resulting in objects apparently bouncing in and out of facilities due to signal inaccuracies. These were improved using a manual investigation of each trace that identified the facility most likely visited. Researchers decided which variations in the data corresponded to actual movement and which were more likely a result of noise.

Inferring the means of transport was more complicated. Due to the long intervals between reports, it was not always possible to say whether a location was submitted while the tracker was moving or resting or how the item was being transported. This could be determined only by considering the whole trace rather than a single report. For example, a report could point to a nondescript facility near a freeway. If other reports were received along the same road, we concluded that the sensor reported en route from a truck rather than from the home. A report from an urban location could have been sent from a residential waste bin, a garbage truck, or a vehicle rushing by on a nearby road. However, the contextual information from a trace of ten to twenty reports allows for a good estimate of the mode of transport and the purpose of the trip.

A second source of public data came from the municipal collection contracts published on the city's website (Seattle Public Utilities 2010). Both Seattle and Washington State keep statistics on the amounts of waste collected and processed by each facility, as well as tipping fees—the disposal fees for "tipping" a truckload of waste at a landfill (Washington State Department of Ecology 2010). These contracts specify the process and destinations per waste type, with the exception of household hazardous and electronic wastes, which play a special role in this study.

### Identifying Visited Facilities in Traces

Based on information from the available sources, it was possible to identify most of the facilities from which the sensors reported. We received reports from landfills, recycling centers, and compost facilities, but also unexpected locations such as residential neighborhoods. In many cases, the report would be sent during movement to or from a facility involved with transportation. Such reports were useful in determining the mode of transport utilized. In some cases, this was evident, such as reports received from shipping terminals on airports, harbor docks, storage facilities, truck stops, or rail terminals. This is, of course, only conveniently possible in areas where spatial information is already available: in publicly accessible aerial images and geocoded facility information.

Facilities that could not be identified were further investigated using publicly available information ranging from company directories to published

**Figure 2.6**
A printer cartridge at the Seattle-Tacoma International Airport. Web application: David Lee. U.S. Geological Survey, USDA Farm Service Agency, reproduced under Google Maps fair-use policy.

materials from waste management companies. By evaluating each report, we constructed a network of facilities, companies, and processes involved in the different waste management processes.

Commodity prices for various kinds of scrap materials and tipping fees at landfills offered clues for why a waste contractor chose a specific process or mode of transportation. These data sets are available from public and private sources. Data published by the city of Seattle specify estimated values for different kinds of curbside recycling items and tipping fees at regional landfills. Spot market prices[7] for commodities were acquired from a commercial database providing real-time commodity price and market trend information for different kinds of recyclable materials (RecycleNet Corporation 2010).

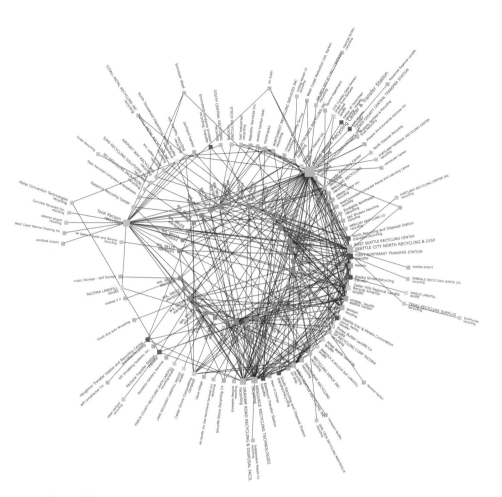

**Figure 2.7**
Network visualization of the material streams and facilities visited by the tracked items (collected in different ZIP code areas, gray in the center). Visualization by the author, courtesy MIT Senseable City Lab, 2010.

## Characterizing Traces by Transportation Distance

Traces that exceeded a distance of 1,500 kilometers were primarily associated with cell phones, printer cartridges, and batteries. A few cell phones reported trajectories to Chicago, Atlanta, and ultimately Florida. Two printer cartridges sent their last report from the same facility just across the Mexican border. What is remarkable about these two traces is that they reported very different routes. One arrived through California by truck, the other, by rail via Chicago, resulting in a difference of several thousand kilometers.

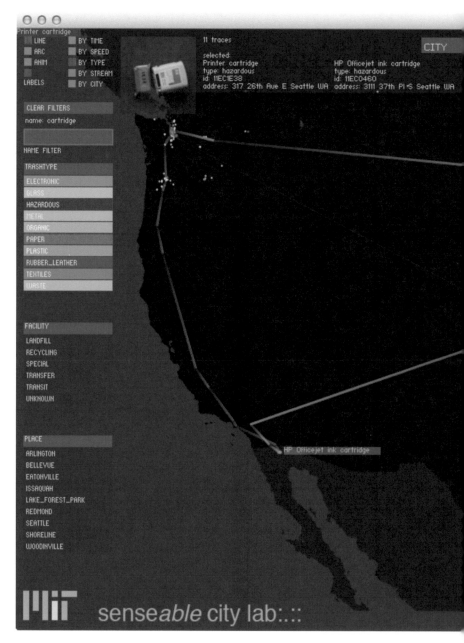

**Figure 2.8**
Visualized traces of two printer cartridges, traveling from Seattle to the California–Mexico border via two different routes: truck through California (red) and train via Chicago (blue). The transport GHG emissions associated with the blue track are lower than the red one. Visualization by the author, courtesy MIT Senseable City Lab, 2010.

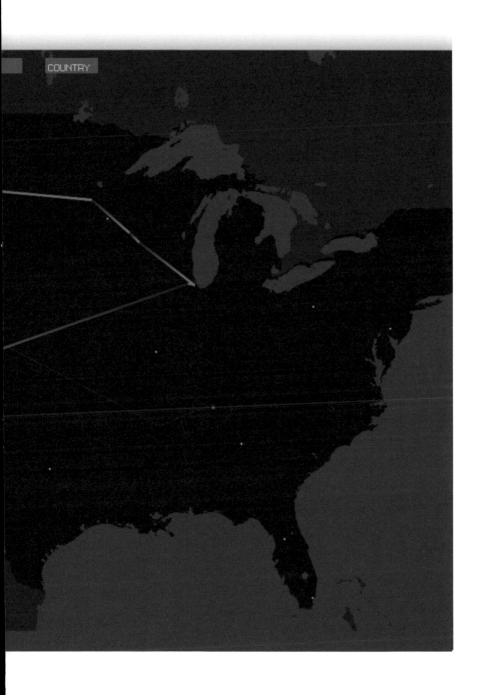

While the sensors were limited in their capacity to send reports from outside the United States, a number of reports were received from the Vancouver region in Canada and from the Mexican border region, which is partly serviced by the U.S. cellular network and its partners. Other reports of paper, plastic, and electronic waste were received from harbor facilities en route to the Pacific Ocean, indicating overseas export.

Each trace can be represented as a sequence of facilities that had been identified. Individual chains can be combined into a network of facilities and waste streams from which disposal, recycling, and special treatment pathways can be inferred. The properties of this network, its central hubs and peripheral areas, indicate the interactions between different companies participating in the waste removal process. Since the Seattle area is divided into different service districts managed by different companies, the visited facility differed by area of collection.

The acquired traces reported from up to four facilities, with the Seattle MRF visited most often. A substantial number of sensors reported their final destinations from different landfills. Specialized recycling facilities for batteries and metal smelters were identified, as well as a wide range of different transportation facilities. The network is not complete, however. Due to the low temporal resolution of the reported trace and possible premature device failure, some facilities are likely missing from the data set. Furthermore, few reports could be acquired from waterways and areas unserved by the cell phone infrastructure. It was nevertheless relatively simple to identify large facilities such as landfills, airports, and train terminals.

## Problems of Inferring Causality

The analysis of the collected GPS traces requires reading on different levels. Due to their sparse structure, the GPS traces offered little explicit information beyond extensive facilities and approximate locations visited. However, by relating and comparing different traces and places of interest, a surprising wealth of clues and implicit information could be gleaned. To some extent, the limits of quantitative analysis are a result of the small sample, but it should not be missed that many fundamental questions of spatial analysis are qualitative in nature, even when addressed through the quantitative proxies of geographical coordinates. Meaning and causality are implicitly inferred through proximity. We assume that an item reporting from a location close to a landfill has been disposed of there; however, such inferences can be problematic. In many cases, different causes for reporting from this location are plausible—for example, a freight rail line passing by the landfill. A GPS trace can rarely count as evidence for a particular activity having taken place. However, it is a clue that warrants further examination.

In the recorded data set, we found some unexpected locations. A few trajectories led from the MRF to a residential neighborhood, which could be explained by workers salvaging objects from the waste stream. Other recorded traces—it is unclear which service collected the corresponding items—sent reports indicating the location of a facility unrelated to waste disposal. Fourteen objects, including typical household recyclables, seemed to have reported from this facility over the course of two weeks. Upon inquiry, the manager of the facility responded by email: "I have no idea what you are referring to. Our [redacted] facility located at [redacted] does not accept paper, aluminum cans, plastics, or e-wastes. I do not know of a landfill that was located near our facility. No garbages or waste dumping were accepted at any time at our facility. We are permitted to accept concrete, asphalt, clean fill, topsoil, sod, stumps, and brush."[8]

This answer might have been perfectly sincere, since there are a wide variety of reasons that could explain the trackers reporting from this location without the knowledge of the facility staff. The items could have arrived at the facility commingled with construction waste that the facility was permitted to receive. The trackers could have become separated from their objects and somehow ended up in the truck used for hauling. This event clearly demonstrates the limitations of GPS data as evidence and the need for follow-up investigations on the ground.

## Contextualizing Growing Waste Distances

During the various stages of the Trash Track experiment, we set up a collection and tagging point, as well as a public display in the Seattle Public Library. Using this installation, the public was able to follow the movements of trash in real time. Participants in the post-experiment study received access to a special website where they could analyze and investigate the traces recorded during the experiment (Lee et al. 2014). Lee and colleagues investigated the effect on volunteers when they saw the recorded data. The researchers found that volunteers who indicated that they had a good sense of where hazardous waste disposal sites were located grew less confident in their knowledge after seeing the data in all their complexity (ibid.).

To a general audience, the map of all recorded traces may demonstrate little more than the complexity of the waste systems and the long distances involved. Waste management professionals read it differently. When I presented the U.S. map of the recorded traces at a recycling conference

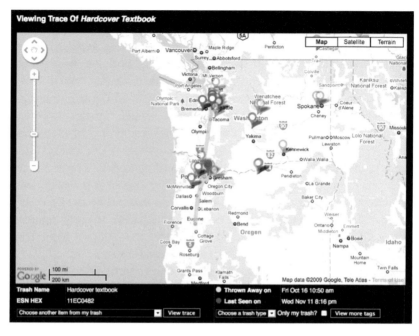

**Figure 2.9**
Website for reviewing traces used in the volunteer study (Lee et al. 2014). Courtesy
MIT Senseable City Lab, 2010.

in Washington State, a member of the audience immediately called attention to an inconspicuous data point and identified it as the location of an
e-waste recycling company he has worked with in the past.

To move beyond the anecdotal and evaluate the environmental impact
of waste transportation, it is important to understand the relationship
between the properties of the discarded objects and their end-of-life transportation distances, the collection mechanism, and the geography of where
the items have been disposed. As of now, little reliable data about waste
transport exist, and public agencies rarely report any transportation-related
statistics in their annual MSW reports.

## The Growing Distance to Landfills

The relatively long transportation distances to landfills observed in the
experiment were expected. Until the late 1970s, cities and larger towns
operated their own landfills. Once groundwater contamination and health
problems stemming from the leachate of unlined dumps were recognized, the 1984 hazardous and solid waste amendments of the Resource

Conservation and Recovery Act (RCRA) mandated replacing unlined dump-sites with sanitary landfills (U.S. Congress 1984). To mitigate damage to the environment and public health, the RCRA required constructive measures for isolating content, leading to higher operating costs and tipping fees. As a result, the number of landfills decreased, and the remaining ones became larger and located farther away in areas with low land value.

Growing waste transportation distances did not mitigate environmental and sociopolitical issues. The siting of large waste facilities continued to raise questions of environmental justice, as the waste streams kept following the path of least resistance to the vicinities of underprivileged communities (United Church of Christ 1987; Bullard 2000; Pellow 2004). According to estimates, around 10 percent of MSW in the United States is disposed of in a different state (Repa 2005, 2), making the interstate waste trade a source of numerous disputes among cities, regions, and states.

## Is Distance a Problem?

In response to concerns about long-distance waste transportation, the European Union's waste policy and the Basel Convention, an international treaty on the control of transboundary movements of hazardous wastes, embraced the Proximity Principle, postulating that waste should be managed at the "nearest appropriate installations" to where it is generated (European Commission 2008; Kummer 1999). However, is it really a problem when waste transportation distances grow? Even from a purely environmental standpoint, it is neither practical nor desirable to process all wastes where they are generated, but rather in facilities that meet the necessary safety standards in places where waste does not harm humans and the environment. A large service provider with state-of-the-art equipment and sufficient regulatory oversight might in this regard be preferable to a network of small waste haulers and local family businesses.

Life-cycle assessment (LCA) analyses seem to confirm that transportation is one of the smaller environmental impacts in terms of emissions. For most recyclable materials generated by households such as paper, cardboard, and metals, the energy savings of recycling easily compensate for the losses generated by collection, transportation, processing, and remanufacturing (Morris 2005). It is, however, not clear how accurately LCA models represent the current realities of increasing distances of waste transportation. The Waste Reduction Model (WARM), an LCA model used by the EPA for estimating greenhouse gas (GHG) emissions of waste systems, assumes a standard transportation distance to a disposal site of twenty miles, or about thirty-two kilometers (Scharfenberg, Pederson, and Choate 2004, 2). Under these assumptions, transportation has a negligible impact on the overall result, and when allowing for triple the amount of transportation energy, overall GHG emissions would rise between 4 and 20 percent depending on the disposed material (ibid., 3). In this context, it is noteworthy to consider that Seattle's main landfill is located more than 400 kilometers from the city. While trash at this particular landfill arrives by rail and therefore generates substantially lower emissions than trucking, such distances are not unusual for contemporary waste systems. WARM does not, however, account for multimodal, long-distance waste transport involving multiple stops, temporary storage facilities, and transfers (ibid., 2).

These conditions, however, were not exactly uncommon in the recorded data, considering that the longest traces collected during the experiment were more than 6,000 kilometers and many traces involved multiple modes including trucks, trains, and planes. Based on the limitations and assumptions of WARM, the role of transportation should therefore not be easily dismissed.

Long transportation distances may have other negative implications. Political scientist Jennifer Clapp describes the psychological effect of "distancing," which obfuscates information and diminishes perceived responsibility. With physical distance, the mental distance between generators and their waste grows as well, which may lead to increased waste generation as awareness of the implications diminishes (Clapp 2002). Long distances also obfuscate any unequal distribution of waste generation and the burdens of its disposal. The pollution haven hypothesis alleges that longer transportation distances and waste exports are associated with lower social and environmental standards at the recipient's end (Smith, Sonnenfeld, and Pellow 2006; Puckett et al. 2002; Puckett et al. 2005).

With waste changing its owners multiple times en route across the globe, the topological distance of waste removal is a relevant factor, especially when data are not universally collected, exchanged between contractors, and shared with the public. The possibility that violations such as illegal dumping or "fly-tipping" of hazardous material out in the wild go unnoticed is not the only concern.

A subtler, more pervasive problem is the likelihood of collected information degrading along the way with longer transportation distances, multiple ownerships, and the absence of universal standards of monitoring and enforcement. Policies to incentivize recycling can give rise to practices such as mislabeling hazardous waste as benign recyclables, diverting the material toward inappropriate processes or abandonment in transit. Even if everything is handled by the book, a lack of information impedes the evaluation of best policies in waste systems.

In recycling, benefits and consequences are difficult to establish. Simple measures such as diversion rates can be misleading, since it is not always clear what happens to collected material. The city of San Francisco reports an exceptionally high diversion rate of 80 percent, meaning that four times more recyclable material is collected than residual waste. However, as MacBride notes, cities typically do not report where and how much of the diverted material is recycled and how much of it is exported. The diversion rate is useful for assessing the quality of a recycling collection system, but it does not account for anything that follows curbside collection (MacBride 2012).

The problems of accounting for recycling practices are often hidden in details such as exemptions and material designations. Labeling of packaging materials, information disclosure, and even the exact meaning of terms such as "recyclable" are highly contested, for example, to reflect whether a material is theoretically recyclable or is actually recycled in the local system.

Deliberate vagueness and ambiguities of designations can be found at almost every corner. The ASTM resin identification codes[9]—the numbers encircled by recycling arrows on the bottom of plastic containers—only inaccurately indicate whether an object can be recycled. Of the wide variety of plastics in use, only a few resin types such as polyethylene terephthalate (PET) and high-density polyethylene (HDPE) can be recycled easily, and then only if clean, dry, and presorted. Materials collected in a single stream from the curbside can be difficult to separate through the automatic mechanical, optical, and magnetic methods used in MRFs, while many packaging designs that use composite materials cannot be recycled at all.

### Evaluating the Environmental Impact of Transportation

The question of whether long transportation distances offset the benefits of recycling depends on many assumptions and values that can only be approximated. Feeding the collected traces into existing LCA models allows both an estimation of environmental impact and an examination of the assumptions underlying the LCA process.

Different modes of transportation require different amounts of energy. To move the same mass, a garbage truck requires tenfold the amount of energy of a freight train—both modes are used in the U.S. waste system. The EPA estimates the total energy consumption for collection and transportation at 345 MJ per ton of material landfilled (U.S. EPA and Office of Air and Radiation 2006, 7.16).[10] Using the values of table 2.2, this would correspond to a transportation distance of 140 kilometers for trucks, or 580 kilometers for rail under ideal conditions. Since the vehicle has to return the same way, this would result in an average landfill distance of seventy kilometers for a truck and 290 for a train, which is shorter than the distance from Seattle to its main landfill. This simplified calculation neglects many additional factors considered in LCA models, which would further shorten these distances.

Based on its chemical composition, one liter of diesel produces a GHG equivalent to 2.68 kg $CO_2$ when burned in a combustion engine (U.S. EPA 2006). Assuming that a fully loaded twenty-ton garbage truck has an average fuel efficiency of 0.4 liters per kilometer (Gaines, Vyas, and Anderson 2006), the truck emits approximately the equivalent of fifty-four grams $CO_2$ per ton and kilometer. In comparison, WARM uses a higher value of seventy-four grams $CO_2$e/ton-km[11] as a basis for calculating the transportation impacts, acknowledging trucks that are not entirely loaded and other efficiency losses.[12] A single garbage truck traveling seventy kilometers to a landfill and back again would therefore produce 150 to 200 kilograms of $CO_2$, which is about half of what a two-person U.S. household generates per week (U.S. EPA Office of Atmospheric Programs 2006).

Using WARM with the measured distances for different types of waste, the GHGs stemming from the transportation involved in the curbside collection of paper, plastic, and metal seem insignificant compared to the overall impact of waste disposal (table 2.3). Paper, cardboard, and especially metal offer high GHG savings when recycled with state-of-the-art processes and have high inherent commodity values. Plastics, due to their wide range of shapes and materials, are a complicated case; the results range from significant savings to almost none. Glass, a cheap and inert material, is also a borderline case. According to WARM, the recycling of glass items offers the

**Table 2.2**

Fuel consumption for different modes of transportation (Davis, Diegel, and Boundy 2009)

| Mode of transport | kilojoules per ton kilometer |
|---|---|
| Class 1 railroads | 246 |
| Domestic waterborne | 370 |
| Heavy trucks | 2,426 |
| Air freight (approx.) | 6,900 |

**Table 2.3**
Recorded distances by waste type and corresponding GHG emissions (tons of $CO_2$ equivalent emissions per ton of material) based on EPA WARM, assuming one-way transportation with a fully loaded garbage truck. Note: shortest distances correspond to sensors destroyed prematurely.

| Trash Type | Mean dist. (km) | Mean GHG (ton $CO_2e$/ton) | Min. dist. (km) | Max. dist. (km) | Max. GHG (ton $CO_2e$/ton) |
|---|---|---|---|---|---|
| printer cartridges | 1713.57 | 0.100 | 1.16 | 6151.71 | 0.358 |
| cell phone | 831.14 | 0.049 | 5.56 | 4825.22 | 0.281 |
| NiCd battery | 1128.47 | 0.066 | 6.62 | 4443.76 | 0.259 |
| alkaline battery | 458.64 | 0.026 | 3.97 | 4374.11 | 0.255 |
| lithium battery | 1246.15 | 0.073 | 4.84 | 3975.58 | 0.231 |
| fluorescent bulb | 313.64 | 0.019 | 3.34 | 3454.86 | 0.202 |
| other plastics | 61.11 | 0.003 | 0.02 | 2814.8 | 0.164 |
| other e-waste | 97.91 | 0.006 | 0.09 | 678.07 | 0.040 |
| books | 75.09 | 0.004 | 0.49 | 616.85 | 0.036 |
| cardboard | 67.31 | 0.004 | 0.02 | 608.02 | 0.035 |
| corr. plastic cup | 76.78 | 0.004 | 1.76 | 529.46 | 0.031 |
| wood | 92.36 | 0.006 | 1.22 | 515.89 | 0.030 |
| glass jars | 50.89 | 0.003 | 1.32 | 488.63 | 0.029 |
| mixed | 71.5 | 0.004 | 0.6 | 481.59 | 0.028 |
| ceramics | 84.49 | 0.004 | 0.82 | 447.13 | 0.026 |
| textiles | 70.48 | 0.004 | 0.41 | 459.17 | 0.026 |
| shoes | 58.96 | 0.003 | 0.21 | 431.88 | 0.025 |
| plastic bags | 58.01 | 0.003 | 0.21 | 380.35 | 0.022 |
| other hazardous waste | 90.14 | 0.006 | 0.52 | 347.29 | 0.020 |
| styrofoam | 46.33 | 0.002 | 0.79 | 294.96 | 0.018 |
| other paper | 49.35 | 0.003 | 1.16 | 306.94 | 0.018 |
| aluminum | 32.86 | 0.002 | 1.29 | 274.39 | 0.017 |
| other corr. plastic | 54.62 | 0.003 | 4.13 | 275.36 | 0.017 |
| steel cans | 28.67 | 0.002 | 1.15 | 281.08 | 0.017 |
| plastic bottles | 27.15 | 0.001 | 0.03 | 283.54 | 0.017 |
| corr. cardboard | 29.49 | 0.002 | 0.8 | 291.05 | 0.017 |
| computers | 101.24 | 0.006 | 0.92 | 269.4 | 0.015 |
| tires | 135.7 | 0.008 | 2.48 | 271.68 | 0.015 |
| scrap metal | 31.98 | 0.002 | 0.98 | 272.49 | 0.015 |

Table 2.3 (continued)

| Trash Type | Mean dist. (km) | Mean GHG (ton $CO_2e$/ton) | Min. dist. (km) | Max. dist. (km) | Max. GHG (ton $CO_2e$/ton) |
|---|---|---|---|---|---|
| CRTs | 49.75 | 0.003 | 5.04 | 239.59 | 0.014 |
| furniture | 79.46 | 0.004 | 4.25 | 248.54 | 0.014 |
| periodicals | 21.95 | 0.001 | 0.9 | 224.03 | 0.013 |
| cartons | 16.39 | 0.001 | 1.04 | 187.6 | 0.011 |
| glass bottles | 18.4 | 0.001 | 2.71 | 84.07 | 0.004 |
| rubber | 11.67 | 0.001 | 3.24 | 34.92 | 0.002 |
| spray cans | 10.94 | 0.001 | 0.93 | 45.1 | 0.002 |

lowest GHG reduction among the materials collected at curbside. The traces collected from tracked glass items had a maximum length of 488 kilometers. This distance would translate to 0.046 tons GHG generated per ton of material counting both directions, which is more than half of the GHG emissions saved by recycling the same amount of material.

For the long traces of household hazardous waste and printer cartridges, the impact of transportation is even more significant. The longest trace associated with a printer cartridge generates 0.3 to 0.8 metric tons of greenhouse gases according to WARM, depending on the mode of transportation. This substantial amount effectively neutralizes the expected benefit of recycling, since WARM estimates the GHG reduction of recycling computer scrap as 0.618 tons. While this is only an estimate based on approximate values, it shows that long distances involving multiple modes of transportation significantly diminish recycling benefits, a fact that is not considered in WARM.

Electronic and household hazardous wastes can only be recycled in specialized facilities, and due to the relatively small volumes compared to other wastes, few such facilities exist. Recycling CRT glass, which contains lead that makes up for roughly 25 percent of its weight, is a costly process, and only a few smelters are certified to conduct it (Shaw Environmental, Inc. 2013). The alternatives to long-distance transportation are worse—due to the rising costs of recycling CRT glass, some recyclers stockpile CRTs in their facilities, creating new environmental hazards (TransparentPlanet 2012). The central role of collection and transportation is also reflected in the fact that electronic and household hazardous wastes amount to roughly 2 percent of the solid waste stream but generate transportation costs that can amount to 80 percent of the total recycling cost for an item (Kang and Schoenung 2005). The recorded traces have shown that transportation mechanisms deserve special attention when evaluating product steward-ship models and e-waste recycling at specialized facilities.

## Discussion of the Trash Track Results

Our experiment has indicated that, in some instances, the GHGs generated by the transportation of material can neutralize or at least diminish the benefits of recycling in terms of GHG savings. This finding has implications for the collection of waste and recyclable materials and highlights the need to pay more attention to transportation.

Currently, cities are evaluating a variety of strategies, such as mail-back or take-back programs involving retailers, collections at transfer stations, or in the case of partly reusable appliances, remanufacturing and reuse pro-grams (Michaelis 1995). All of these strategies have different implications for transportation and its impact, but the exact differences are difficult to measure. Since no single strategy, whether centralized collection or relying on citizen participation, is clearly superior to another, evaluation of the actual distances across systems and institutional domains is necessary for a more comprehensive comparison (Norton-Arnold & Co., URS Corp., and Herrera, Inc. 2007).

### Toxic Waste Is Shipped the Longest Distance

It is again worth noting that, based on the results, it seems that the more toxic the material, the longer the transportation distance. During the study, household hazardous and electronic waste items reported the lon-gest traces, due to the number and geographic distribution of specialized treatment facilities. For toxic materials, transport to a distant, but adequate treatment option is certainly preferable to inadequate disposal nearby. It is, nevertheless, worth investigating the means of transportation. Mail-back and take-back programs have many advantages. Since the items are already source-separated by the consumer or retailer, the likelihood of the objects receiving appropriate treatment is higher than in commingled collection. Mail-back and take-back programs are also convenient, if there are enough potential collection points.

As shown in the results of the experiment, however, these hybrid mecha-nisms involving different modes of transportation have the disadvantage of outsourcing a significant part of waste transportation to courier services such as UPS or FedEx. These services are not designed for handling waste and involve potentially more handling, more packaging than necessary, and mode changes, including airfreight. As an example, the IT infrastructure and printer servers are often configured to automatically order new toner cartridges as sensors determine that supplies have run low in a device. The mail-back of single empty toner cartridges in bubble-wrap envelopes seems wasteful, but might make sense if these cartridges can be easily remanufactured and are otherwise harmful if disposed of in the trash. In the collected traces, however, we saw a number of instances of hazardous and electronic wastes shipped by airfreight, which is both environmentally and economically expensive compared to the expected energy savings of recycling these products.

Importantly, the fate and trajectories of electronic and household hazardous waste shipped by couriers largely remain undocumented, as they are no longer part of the waste system. As a result, reliable data about waste transportation seldom exist beyond the first stop—the collection point or the MRF. Given the lack of information exchange and compatible categories between waste haulers and courier services, the environmental impact of waste transportation can only be estimated but not measured.

### Location of Disposal Affects Diversion Rates

Considering the limitation of the small area of the experiment with a handful of municipalities outside Seattle proper, our results indicate that the geography of recycling is not even. Rural municipalities serviced by different companies and public works departments have significantly different recovery rates. Whether an item is recycled or ends up in a landfill can depend on the location of its disposal. Despite its small sample size, the study demonstrated that the odds of an item going to a landfill are significantly higher in some areas than others. Beyond urban and suburban locations in the Seattle metropolitan area, we also deployed some trackers in rural communities a few hours outside of Seattle. Items discarded in these rural areas showed higher odds of ending up in a landfill. The small sample used in the Trash Track experiment did not allow the comparison of cities and districts beyond the anecdotal level: exploring the differences within the geography of waste removal across different regions would be an important area for future study.

### Conclusion: Reading Systems from the Outside

Cognitive scientist Colin Ellard distinguishes between two distinct kinds of maps that can be used to make sense of an unknown territory. The first is defined by a single overriding principle of localization, such as the gradient of the earth's magnetic field. In contrast to this gradient map, the second type is constructed from diverse sources and clues, a hybrid collection of landmarks and heuristic devices (Ellard 2009). The GPS traces collected during the experiment may seem like a prime example of the first category, but as we have seen, even from a technical perspective this is not the case. Localization involves an array of different methods beyond the triangulation of satellite signals, giving rise to unexpected artifacts. Furthermore, to make this map useful, it is important to connect it with the diverse range of other data sources.

There is no privileged position with a full view of the entire waste system. Every actor in the system is, in a sense, an outsider who perceives and represents the system as a different problem. Public information represents waste systems as the proverbial black box. Agencies collect data about inputs and outputs, the amounts of material collected, and the amounts processed at disposal facilities.

Even when inputs and outputs are known, the pathways connecting them are not always clear. In many cases, published hauling contracts remain unspecific by, for example, listing "Asia" as a destination. Although facilities that transport, process, and dispose of waste are regularly monitored by the EPA, the contents of the Facility Registry System (FRS) database are often imprecise, and the classifications of facilities are not consistent across the country.

The method of waste tracking developed during the Trash Track experiment could have a profound impact on waste system policies and practices at different levels. It offers a way to read waste systems from the outside, using GPS traces to connect the dots. Although location sensing allows us to map pathways, integrating various representations of the waste system requires the addition of public information and the context gathered from the actors in the waste system.

From an analytical perspective, the study helped to evaluate the efficiency of removal systems and waste stewardship concepts. From a narrative perspective, the project brought volunteers and visitors to the Seattle Public Library exhibition and closer to their urban services, conveying a sense of the complexity of waste systems and provoking questions about the future of waste policy and practice.

The Trash Track project was too small to offer a generalizable quantitative analysis of the entire Seattle waste system, but it points toward a worthwhile effort to conduct a similar experiment under narrower constraints, focusing on a specific pathway or collection mechanism.

Because location traces represent actual processes, even anecdotal findings offer insights. By documenting the global scope of waste removal and its various modes of transportation, the collected data can support the evaluation of waste policies and inform decisions on the proper treatment of material. In the end, the Trash Track project demonstrated how the reality of the formally planned and managed municipal waste system is more complicated than its explicitly prescribed procedures, contracts, and representations might suggest. The following epilogue discusses the implications of the experimental method of waste tracking for different actors and their perspectives on the waste system—policy makers, companies, citizen scientists, or watchdog organizations.

# Epilogue to Part I: Waste Forensics

The Trash Track study established the legibility of waste systems through a novel approach of data collection and a real-time visual representation of the state and flows of the waste stream. Attaching location trackers to individual items of garbage is a method open to almost anyone, including watchdog organizations, municipalities, and waste service providers. It allows one to estimate the performance of recycling, supports the investigation of environmental crimes, and offer ways to monitor a waste system without access to internal information. The collected data, however, do not speak for themselves and require contextualization with other sources to become meaningful.

Since the completion of the study, its experimental methods have garnered interest from diverse groups and actors involved in monitoring waste systems. Local governments seek to evaluate their policies and engage the public in an educational experiment. Companies want to improve their reverse logistics. Watchdog organizations and environmental activists hope to find new tools for conducting investigations.

What can a tracking method that in principle does not require privileged access beyond costs and labor mean for the legibility and governance of waste systems? Which purposes and groups can the method serve? In this epilogue to part I, I discuss scenarios regarding how data and evidence collected by active location sensing can be used in public discourse about waste policies, including legal controversies and the investigation of environmental crimes. These include social accountability initiatives, environmental crime investigations, voluntary certification programs, and policy tools for local governments.

## Tracking as a Social Accountability Tool

Perhaps the most controversial issue raised by this kind of study is the international trade of hazardous waste and defunct electronic equipment.

In this respect, environmental laws contain an inherent conflict of objectives. On the one hand, laws were written with the intent to encourage recycling. But these incentives can have the paradoxical effect of making violations less enforceable. "Recycling is the password for shipping things to other countries," notes environmental health and justice advocate Jim Puckett, cofounder of the Basel Action Network (BAN). In his estimation, waste exports labeled as recycling operations can lead to environmental and social damages in other countries, which may be worse than the disposal options available in the United States.

A complex system of incentives and legal mechanisms is necessary since recycling typically involves more complex logistics and processes than waste disposal does. However, these mechanisms for promoting recycling can be misused, for example, by relabeling electronic scrap as functioning devices and exporting them as "donations" for later reuse. The temporary storage of CRT glass designated for recycling can often extend to more than a decade, with warehouses full of accumulated material before the stockpile is eventually abandoned and the recycling company closed (Powell 2013).

Existing monitoring mechanisms are not sufficient to prevent these kinds of practices. Since the United States did not ratify the Basel Convention on the Control of Transboundary Movements of Hazardous Wastes and Their Disposal, no mechanisms have been created to estimate how much of such wastes is exported. Often used as a proxy, the Harmonized Tariff Schedule (HTS), the central database for tracking and classifying goods that enter and leave the United States, is not ideal for measuring waste exports, since its classifications do not consistently distinguish between waste and nonwaste. Designed to determine customs duties, the HTS captures traded goods in their monetary value rather than volume and material.

Due to this lack of data, NGOs that advocate against waste exports have difficulty collecting evidence to support their arguments. Conducting anonymous surveys among recyclers is not effective, since it is unlikely that waste exporters will admit to a practice that is universally seen as harmful. Watchdog organizations often conduct their own investigations, collecting anecdotal evidence for the practices and impacts of hazardous waste export. In the case of BAN, this means following shipping containers to Asia or Africa, documenting the environment in which the exported waste is processed, and collecting evidence for the origin of the waste, such as photos of asset labels or clues such as the shape of power plugs. Compared to this time- and labor-intensive form of investigation, sensor deployments are easier to scale. They cannot replace the former, but offer clues and indications to make it more targeted.

As the Trash Track study indicated, linking the collected traces with additional data sources is central to the success of an operation. Pursuing containers can be a simple task if shipping numbers for each container are known. The Port Import/Export Reporting Service (PIERS) database contains bills of lading in anonymized form, making it possible to track container movement up to six months back. Since all shipping operations use outdoor loading docks, an experienced investigator can record container numbers at the facility location.

Because U.S. national law does not provide many options for citizens to file lawsuits against exporters of hazardous waste, BAN has to rely on the court of public opinion. For this purpose, even anecdotal information can be valuable, so long as it presents evidence of significant environmental and social harm. While quantifying the larger systemic effect is often impossible, reconstructing the chain of events in a single case can be very effective for advocacy work.

## Tracking in Law Enforcement

Investigations by watchdog groups and citizen initiatives have limits. Citizen-collected data are rarely accepted as evidence in court, and enforcement agencies have to rely on their own data collection. Citizen action may spur an investigation, however, providing background information or a blueprint for additional collection efforts by a law enforcement agency.

National and international agencies investigating environmental crimes, such as the U.S. EPA or Interpol, face similar challenges of sparse and inconclusive data. U.S. waste monitoring relies mainly on self-reporting, which inescapably involves a level of fraud. According to a prosecutor of international environmental crime, public agencies cannot investigate each suspicious case at a significant scale beyond creating symbolic deterrents due to a lack of resources. However, the prosecution of environmental crime is limited by the intricacies of environmental policy more than by diminishing budgets. This starts with the definition of waste—the U.S. Resource Conservation and Recovery Act requires proof that a material is actually hazardous waste, which makes practices of mislabeling particularly difficult to address. In court, this burden lends weight to defense rebuttals such as, "The CRTs looked fine when we put them into the container." Proving a violation requires additional evidence such as showing that the price paid for a service is consistent with the price of dumping. Due to these complexities, a single case can take up much of an agency's resources.

Also due to such constraints, the comprehensive monitoring of waste exports with self-reporting sensors in sample sizes large enough to find a "needle in the haystack" is beyond the reach of federal and international agencies. Nevertheless, location tracking is commonly used in law enforcement as both evidence and an indicator for further investigation. The enforcers typically rely on whistleblowers (such as truck drivers documenting suspicious container numbers) to initiate investigations, especially when multiple parties conspire to commit fraud. The relationship between citizen initiatives and enforcement agencies is not always without friction. Complaints abound that data collected and observations reported by citizen scientists are ignored by federal agencies.

### Tracking for Voluntary Monitoring Programs

In the absence of stringent regulation of waste exports, voluntary stewardship programs have emerged. These programs, implemented by industry or NGOs, provide certifications for recyclers who comply with regulations and follow best practices. The E-Stewards certification program implemented by BAN requires recyclers to demonstrate that they do not engage in waste exports. Being subjected to such additional scrutiny can be lucrative for recyclers: it gives them access to high-profile clients that want to eliminate the risk of their waste showing up in the wrong places. However, certification programs also can be gamed, for example, by keeping double books for yearly audits. To address such concerns, unannounced site visits and occasional waste-tracking experiments could become part of the certification process. A second group of industry-monitoring programs addresses the problem that regulatory compliance is not enough to insure a company against legal action. The enforcement of environmental violations against RCRA is based on the concept of strict liability, making companies liable for the actions of their waste management contractors, regardless of knowledge or intentions.

Beyond certification programs, some companies have their own supply-chain forensics programs to investigate hidden potential issues in their supply chain and waste management arrangements (Deloitte 2015). In a particularly egregious case, a Singapore-based electronics recycler that was contracted by a chip manufacturer to destroy and recycle faulty computer chips exported these chips to Hong Kong and resold them on the black market as new products (Vijayan 2008). The waste stewardship program operated by CHWMEG,[1] a non-profit trade association of industrial manu-facturers, is aimed at helping companies assess and improve their regula-tory compliance by offering environmental health and safety audits of their facilities. Unlike the E-Stewards program, CHWMEG keeps findings confi-dential and excludes NGOs and governmental organizations from member-ship. Taking a nonadvocacy stance, it does not interfere with enforcement or publicly approve and certify contractors. Instead, CHWMEG indicates whether a facility has been reviewed in its public database.

From the environmental policy perspective, voluntary programs can be a mixed blessing, as described by a federal prosecutor. Certification programs reward companies for holding themselves to standards more stringent than environmental laws, but skepticism about "greenwashing" remains. Some enterprises that have been convicted of environmental crimes had certifi-cations and audits on their books. Nevertheless, the participation of the recycling industry is indispensable in implementing environmental poli-cies that are too complex to be monitored and enforced by a single cen-tral agency. A compromise between command-and-control and voluntary approaches are mandated certification programs that are implemented and controlled by an independent third-party agent, as currently realized in the LEED standard for green building construction.

### Tracking as an Evaluation and Education Tool for Municipal Services

Local governments can conduct waste-tracking studies to calibrate their recycling systems by defining collection mechanisms and treatment options for specific materials. An overly ambitious recycling program that mandates recycling for too many materials can backfire if the municipal system cannot handle the proper management and treatment of these materials. Seattle, for example, had initially banned a long list of items from household trash and mandated special procedures to add them to the recycling stream, which simultaneously confuses residents and exceeds the capabilities of local recycling options.

Operation Green Fence in China highlighted problems in U.S. recycling systems that may require such recalibration (Recycling Today 2013). In April 2013, the Chinese government began random inspections of imported recyclables at ports of entry, rejecting whole shiploads if a single bale of recyclable material exceeded levels of contamination with other materi-als defined in Chinese regulations. The operation affected the business of many U.S. recyclers and raised questions about whether single-stream or source-separated collection models are more effective for raising the quality of recyclable materials.

In my interviews, both recyclers and local government representatives claimed that the behavior of citizens and businesses plays a central role in the performance of a recycling system. People have to learn how the system works, know whether to leave the items on the curbside or take them to a drop-off point, and understand how to prepare materials since contaminants like food waste turn recyclables into garbage. Under this rubric of engagement and education, public waste-tracking experiments do not merely explain how the waste system works, but instead show it in action. The traces produced by trackers offer clues that instigate questions instead of supplying finished narratives.

# II   Informality

# Prologue to Part II: Making Informal Waste Systems Legible

When archaeologist Bill Rathje and his team of "garbologists" excavated landfills to study contemporary culture, they developed methods to quantify and classify discarded scraps, packaging materials, and newspapers. They determined the age of landfill strata using a taxonomy of beer can ring-tabs, which, until manufacturers replaced them with nondetachable rings, varied in design according to manufacturer and production date (Rathje and Murphy 2001).

Surprisingly, they found that the results of consumer behavior surveys did not match the collected evidence. What people say they throw away turns out to be different from what they actually throw away. These findings were often counterintuitive. Rising meat prices, for example, led people to purchase more meat, leading Rathje to conjecture that people tend to stockpile items when prices go up, but they fail to consume and preserve their purchases.

The experiment conducted by Rathje's team shows how data collected from waste touches on the most private aspects of residents' lives—material circumstances, ages, preferences and social codes, health, and even pets. Similar landfill excavations have played a role in writing a chapter of video game history by unearthing a batch of rare game cartridges from a site in New Mexico where the company Atari buried its unsold products (Guins 2014). They have offered insights into the musical tastes of the Grateful Dead's Chosen Family commune in California (Parkman 2014). The term "garbology" has also been claimed by gonzo-journalist Al Weberman, who retrieved garbage bags from Bob Dylan's Greenwich Village townhouse and publicized his analysis of their contents (Weberman 1980).

Like garbologists, cities and waste management firms periodically separate a representative truckload of waste, sort and classify the discards, and measure humidity and weight, paying particular attention to materials such as asbestos or electronic waste. The Spanish art collective Basurama

**Figure II.1**
View of the COOPAMARE facility, located under viaduct Paul VI in Pinheiros/São
Paulo, November 2011, photo by the author.

conducts such waste audits as public events, where participants are invited
to help separate and classify two tons of garbage. Similar information about
waste composition is generated as a byproduct of sorting procedures in
recovery facilities.

Despite their different purposes and scope, the practices of the artist
Olbrich, the archaeologist Rathje, the activist Weberman, and the waste
composition studies conducted by local governments share many similarities. However, only the latter are a formal part of the waste system.

## Legibility through Formalization and Vice Versa

So far in this book, I have emphasized the lack of legibility of waste systems
as a way to reverse-engineer an existing, formally planned system. In part
II, I focus on informal systems and how the different actors involved in
these systems produce legibility, and how these practices of legibility affect
the services and practices they are designed to monitor. The following case
study reflects upon the results of a participatory design experiment with the
goal of developing a system for documenting the routes and activities of
informal waste collection and recycling. In the context of efforts by Brazilian governments to formalize waste pickers, this section argues that formalization not only depends on new systems of legibility, but that producing
legibility is an act of formalization in itself.

Brazil has a history of waste picker unionization and a strong cooperative movement, making the country a hotbed for experiments integrating
waste pickers into the formal waste management system (Medina 2010).
In 2010, Brazil was the first country to pass national legislation that recognizes the profession of waste pickers and formally integrated them into
the waste management system. While promising new opportunities, the
legal recognition of waste pickers comes with obligations of professionalization, including monitoring and data collection requirements that recycling
cooperatives often struggle to meet.

## The Forage Tracker Experiment

The Trash Track experiment we conducted in Seattle demonstrated how
even a modern, integrated waste management approach leaves a level of
uncertainty about the fate of discarded materials. If a modern waste system
turns out not to be entirely legible, one wonders what a similar experiment would reveal in places where waste management happens in mostly
informal ways. In the Forage Tracker experiment described in chapter 4,

my team from MIT used location-based technologies to analyze practices in Brazil's informal system of recycling.

The idea began when a colleague from the University of São Paulo, Professor Maria Cecilia Loschiavo dos Santos, philosopher, and advocate for a local waste picker cooperative, expressed a particular concern. She feared that local governments and policy makers were not aware of the full extent of services that the co-ops provided for their communities due to the invisible and tacit nature of their work. Together we hypothesized that mapping material collection along with spatial decisions and service areas would make the cooperatives more visible and benefit their negotiations with other actors in the waste sector.

The Forage Tracker project was conducted between 2011 and 2014 by Senseable City Lab researcher David Lee and myself in partnership with the University of São Paulo, the Fundação Joaquim Nabuco, and other local partners. The project shared many similarities with Trash Track, as it also involved tracing the movement of discarded materials through the waste system. In this case, however, the routes of waste were not determined by logistic schemes, but by the decisions and experience of individuals. Understanding these decisions and their implications was the central concern. Unlike the Trash Track experiment in Seattle in which location sensing served as a method of passive observation, Forage Tracker, therefore, relied on location-based media as a method for active documentation, in which the workers of the cooperative are in charge. If Trash Track stands for approaches that use large-scale deployments of urban sensors to collect generic data, then Forage Tracker represents smaller, artisanal approaches to data collection that are concerned with the local scale and tailored for a particular group of users. Forage Tracker is one among several projects that have chosen such an approach. Later work that follows a similar approach in the informal waste space includes the Wecyclers project in Nigeria that uses local data to manage a recycling collection system with a fleet of cargo bikes, or Kabadiwalla Connect, an Indian initiative using smartphone apps to map material sources for informal collectors and scrap-shop owners. In all of these initiatives, the amount of data collected and their generalizability and robustness are secondary to the concern about how data correspond with the local practices of the people who use them.

# 3 Local Legibility

It is said that all garbage in Mexico City goes through seven stages of sifting. From the moment it is left on the street until its final destination at the garbage dump on the outskirts of the city. On the night of 4 February 1994, I placed seven small bronze sculptures, painted in seven distinct colours, in seven different garbage bags. I then dropped the bags in garbage piles in seven different districts of the city. In the following days, months, years, I have wandered the flea markets in the city looking for the sculptures to resurface. To date I have found two of the seven.
—Francis Alÿs about his art project "The Seven Lives of Garbage" (Ferguson, Alÿs, and Philbin 2008, 68)

## Informal Waste Management

Waste has a social life. Objects often go through many hands before they are thrown away. They may be passed on in families and between individuals, modified and repaired in the process. When they are discarded, they may be recovered from landfills and dumpsites by waste pickers, sorted, sold, and resold. All of these processes of use and reuse take place outside of the formal waste system.

Municipal solid waste systems of developed countries were established with the intention of replacing the informal economies of scavengers and scrap dealers, who were often seen by urban elites as nuisances and competitors for generating revenue from saleable refuse (Strasser 1999, 114). Despite these efforts, informal waste management continues to play a significant role in many parts of the world. Scavenging and the valorization of collected materials constitute a substantial part of waste management in the cities of both emerging and industrialized countries (Medina 2007b; Simpson-Herbert 2005).

While recycling in rich countries is a technology-intensive industry, the majority of recycling activities in cities of the Global South are carried out by informal networks of waste pickers and scrap dealers. Their value chains involve many stages of selection, sorting, aggregation, and processing that increase the value of the reclaimed materials at every step.

Until recently, informal activities such as waste picking and street vending were seen as temporary phenomena bound to disappear in the course of modernization and industrialization. Yet the formal waste management and recycling industry did not succeed as expected in developing countries. Today it has become clear that the informal economy is here to stay. Although still illegal in many places, informal waste management has persisted, leading many policy makers to reassess their opinions and move from a strategy of criminalization to one that recognizes waste picking as a legitimate occupation that delivers a public service and contributes substantially to resource conservation and sustainability.

An inclusive approach to recycling integrates the informal sector with the formal waste system. This begins by recognizing informal recycling practices and by supporting organizations that allow waste pickers to work under better conditions. Inclusive recycling can involve the public sector hiring waste pickers, offering healthcare and education, providing access to loans and technical support, and encouraging waste picker unionization and the establishment of cooperatives and associations (Scheinberg 2012).

In several countries, waste pickers have formed larger groups that allow them to pool material and sell it in larger quantities, commanding better prices than they could receive individually. In 2010, Brazil enacted its National Policy on Solid Waste (NPSW), the world's first national legislation that recognizes the work of waste pickers as a profession (Brazil 2010). Written with the participation of waste picker organizations, the law grants pickers a central role in waste management and resource recovery. It encourages the formation of waste picker cooperatives and associations, requiring companies and public institutions to form partnerships with them.

There is little disagreement about the goals of waste picker integration—giving workers access to a safe environment that enables economic survival under dignified conditions. The details of this integration between formal and informal actors are unsettled, however, especially since many cities know very little about their formal waste system, if we are to judge by the available data, which can be used as a proxy for estimating the quality of a city's waste system and governance (Wilson et al. 2012, 253).

Informal waste management has been studied through different macro and micro lenses by estimating the economic impact of the sector and publishing ethnographies of waste picker communities. Between those two perspectives, a substantial gap remains. The spatial organization of informal activities on the urban scale as well as the geographies of collection and valorization networks and their links to the global economy remain poorly understood. At the same time, this is the scale where most development projects and initiatives operate.

Informal systems are illegible by definition. Individual waste pickers might benefit to some extent when this illegibility protects their sources, clients, and strategies from competing actors and repressive administrations. The situation is different for waste picker cooperatives when illegibility hinders their development and complicates their integration into formal waste systems. Waste picker unions and recycling cooperatives suffer from a lack of information that can allow them to plan their operations more effectively, negotiate better contracts, and increase the public visibility of their work.

Formalization is not only a question of policy and professionalization, however. Observing and describing the informal sector is an act of formalization in itself. Depending on the approach, formalization can have positive or negative consequences for waste pickers. It can benefit them by giving them access to services and protection, but it can also make them more vulnerable by exposing them to competition and regulatory scrutiny or, in the worst case, by displacing them or removing their sources of income. The questions about how to make an informal systems legible, which I discuss in more detail in the following chapter using examples from waste picker cooperatives in Brazil, cannot be separated from the larger issue of what form of integration is desirable. What should be formalized and what should be left unmeasured?

## Why Did Industrial Waste Management Fail in Developing Countries?

Current efforts to formalize waste picking in developing countries were preceded by similar processes that took place in the United States and Europe at the turn of the twentieth century. After municipal water and sanitation systems helped get epidemics and other public health issues largely under control at the end of the nineteenth century, the focus of waste management shifted to the technical problem of organizing waste collection and disposal at a large scale (Melosi 2004, 50). In 1901 sanitary engineer M. N. Baker noted that "it is seldom recognized that the problems incident to final disposal are largely engineering in character and therefore should be entrusted to engineers" (Baker 1901).

The comprehensive approach to waste management and recycling taken by municipalities was driven by resource value. U.S. households widely practiced the separation of discarded materials during the late nineteenth century (Strasser 1999). In 1898, New York's Street Commissioner George Waring introduced a comprehensive collection and recycling system that used three bins for organic waste, ashes, and residual garbage. The garbage was separated in a new sorting plant that allowed detailed monitoring of all material streams and successfully captured 37 percent of the collected refuse (Hering and Greeley 1921, 300). The motivation for this new form of industrialized recycling was purely economic. It was meant to cover the costs of municipal public services. In the words of Hering and Greenley, waste recovery is only justified "when, after previous disinfection, it shows a sufficient margin of profit" (ibid., 310). Today, this economic calculation has to consider the increasing costs of disposal in the form of transportation, tipping fees at landfills and waste-to-energy plants, and treatment costs for incinerator ash.

Although resource value is an important driver for recycling in low-income countries, industrial-scale resource recovery has not lived up to its expectations in the cities of the Global South. This has happened for several reasons. Rapid urbanization and a weak tax base have caused many municipalities to struggle with the provision of essential services such as water, electricity, waste collection, and sanitation. The available public services are often limited to affluent areas, excluding or underserving many poorer areas. In addition, informal or low-income neighborhoods with narrow, unpaved streets do not support efficient truck collection, further exacerbating the unequal service distribution.

Waste pickers can compensate to some extent for the deficiencies of public services, but municipalities often forgo this opportunity by excluding them from service provision. In 2002, the city of Cairo attempted to modernize its waste system radically by entering into service contracts with three multinational waste companies. This tactic was meant to replace the traditional door-to-door collection system operated by local Coptic-Christian *Zabaleen* waste picker communities, who collected mixed household waste and fed organic components to their livestock, which city officials had deemed unsanitary and backward (Gauch 2003).

However, after a single year of operation it became clear that the new system has not led to expected improvements, but instead to numerous resident complaints about service fees, poor service quality, and waste accumulating on the streets. While the informal Zabaleen system achieved recycling rates of 80 percent, this number dropped to 20 percent after privatization, partially due to toothless service contracts that do not require higher rates and have limited enforcement options (Fahmi and Sutton 2013, 168). Following failed attempts to achieve a symbiotic relationship between the waste pickers and the collection companies, the government is currently prepared to give the Zabaleen official status and direct service contracts once the contracts with the multinationals expire (Guénard 2013).

### Development Drivers for Local Waste Systems

Besides the value of reclaimed resources, several other factors drive the development of local waste management systems in low-income countries (Wilson 2007). Public health and environmental protection are important drivers in environments that have uncontrolled, illegal dumps and litter on the streets that can clog sewers and cause flooding. If no employment alternatives are offered, the closing of open dumpsites creates a dilemma for waste picker formalization by removing access to the material and income the pickers rely on. Such conflicts have become increasingly common as emerging countries have reorganized their waste systems by delegating responsibility for regulating and implementing comprehensive resource management to municipalities, regions, and national governments. In the political processes of shaping these waste policies, waste picker organizations have become effective advocates in several countries.

The experiments with industrial recycling schemes in Cairo and other cities of the Global South have demonstrated that neither a comprehensive modernization approach nor a system carried by waste pickers offers realistic solutions to resource management in developing countries. Waste picker integration seems to be a necessary element of waste systems that are both equitable and effective.

### How Does the Informal Value Chain Work?

Due to the illegibility of informal waste management, public attitudes toward waste picking are subject to several myths that continue to shape waste policies. Contrary to widespread perceptions, scavengers are not always the poorest of the poor, informal recycling is not inherently disorganized, and waste picking does not play an insignificant economic role (Medina 2007a).

Various models and spatial organizations of informal waste management exist; they vary by region and depend on local culture and economy. Waste pickers collect material from open dumpsites, from the street, or by collecting door to door from residents. They typically sell to local scrap dealers who are frequently informal actors themselves. Itinerant buyers who purchase material from residents combine aspects of waste pickers and scrap dealers. Waste pickers specialize in different materials. Solitary waste pickers without access to spaces where they can sort and aggregate material often focus on aluminum cans and metals, which combine light weight with high value, or they collect paper and cardboard with the help of a handcart. They usually sell their material on the same day they collect it. Other materials of interest include recyclables such as glass, PET and other plastics, commercial and industrial waste materials such as rubber, mixed household waste including food waste, and, in the case of manual scavengers in India, even human and animal excrement sold as fertilizer (Rāmasvāmi 2005).

### Recycling versus Valorization

The term "recycling" describes the separation of valuable materials from residual waste but typically does not account for how value is generated. In comparison, the concept of "valorization" is more useful in the context of informal waste management since it focuses on the commercial value that is generated at each stage of extracting, collecting, storing, and processing (Scheinberg, Wilson, and Rodic 2010). In industrial-scale recycling, the sale of reclaimed materials is a substantial source of income for the recycler but still secondary to the income from municipal contracts. For waste pickers, though, this revenue is essential.

Valorization involves many actors who are connected in a network of interdependencies often described as the value chain, a hierarchy of value creation with waste pickers at its base. Pickers collect material and sell it to intermediaries. These middlemen, sometimes formal companies, at other times informal junk shops, aggregate and sell materials to industrial buyers. Manufacturers need PET and cardboard as secondary raw materials, but they usually buy only in quantities that no individual waste picker or informal scrap dealer can supply. Informal junk shops often do not have enough space and financial resources to pool large amounts of material and wait for the best price. Therefore, they have to sell it to other intermediaries who further aggregate and process it by, for example, cleaning and shredding plastics into granular pellets. Access to a shared space for sorting, processing, and storage is essential for waste picker organizations, since it allows them to bypass middlemen and gain access to customers in the higher ranks of the value chain. Such spaces are often located under highways or in other residual urban spaces.

The waste value chain is dynamic. Prices for secondary raw materials are constantly in flux and multiple process chains exist for different materials.

Value chains are influenced by the availability of material, potential buyers, and the capacity to process recovered materials. A material that offers a good source of income at one point can quickly drop in price and lose significance for waste pickers. The prices of PET and plastics are especially sensitive to regional and temporal fluctuations.

## Theories of Formalization

Formalization is a broad concept that can include measures such as registration and taxation, social security, and legal representation. It can be inclusive or repressive in its intention—integrating or banning informal practices. Early studies were often characterized by a pessimistic tone, depicting waste pickers as victims without agency, while more recent work describes informal recycling as a legitimate profession with many potentials for positive development (Gerxhani 2004). Development economist Martha Chen groups the literature on informality into four main schools: *dualist, structuralist, legalist,* and *voluntarist* (Chen 2012).

Based on work by anthropologist Keith Hart, the dualist model describes the informal sector as a result of the different speeds at which population and the formal economic sector grow. The informal economy is described as an economically significant sector, largely autonomous and with limited interconnections to the formal economy (International Labour Office 1972). The structuralist view introduced by Manuel Castells and Alejandro Portes challenges this notion of autonomy, arguing that the informal economy is closely intertwined with the formal sector, which produces informality through deregulation, globalization, and outsourcing (Castells and Portes 1989). Introduced in Hernando de Soto's work on Peru's informal economy, the legalist view hypothesizes that bureaucratic obstacles and their associated costs for the individual drive workers into informality or prevent informal operations from acquiring a formal status (de Soto 1989). Finally, the voluntarist view argues that workers may sometimes voluntarily choose informality over formal employment due to better economic opportunities and more autonomy (Maloney 2004).

The concept of informality, its policy implications, and the ideological alignments of the four schools of thought diverge. While the dualist view places informality outside of the formal economy, the structuralist view locates informality inside formal systems. Structuralists identify capitalism and market liberalization as drivers of informality. Legalists see them as the solution. Structuralists emphasize the dependency of informal wage earners from corporations or other informal firms, while voluntarists describe a class of independent entrepreneurs who prefer informal work to formal status (Chen 2012).

### Formalization Models

These theoretical models of informality have implications for informal waste policies. Anne Scheinberg groups prevalent formalization models into four different categories that imply different dependencies and relationships of accountability for waste pickers, offer different advantages, and carry their own risks (2010, 2012). In the *service model*, municipalities maintain responsibility for planning and monitoring, paying waste pickers to perform the collection. The *commodities model* aims to turn waste picker organizations into micro-enterprises that accept contracts from municipalities and private companies, generating income by selling the collected material. Between those two options, there are different *hybrid models* in which the city and the waste pickers share responsibilities and revenue from selling recyclables, with the city going beyond recognition and offering active support. Finally, *Community-based Enterprise (CBE)* models involve multiple actors in urban service provision, including residents, NGOs, waste picker cooperatives, and private companies.

In international development, the current emphasis on coalitions and partnerships, official recognition, and tight integration into formal waste systems can make it easy to forget that waste pickers remain vulnerable compared to formal actors such as companies and municipalities. The early observations by Christopher Birkbeck about the informal paper scavengers of Cali, Colombia, are still relevant in this context:

We cannot argue that they should be incorporated into the industrial sector of the economy since they are already part of it. Neither can we argue for increasing their share of the income generated by recuperation in anything but a limited way because of the structural constraints that operate in determining income. The garbage picker may work hard, may have a shrewd eye for saleable materials, may search long for the right buyer; in short, he may be the near perfect example of the enterprising individual. It will not get him far. (Birkbeck 1979, 182)

As Birkbeck, as well as Castells and Portes's conceptualizations of the informal economy suggest, informality is not something that takes place outside of formal systems and is limited to waste pickers and street vendors. Informality can also be found in the actions of planners and public servants. The language of waste regulation is a different subject from its implementation and enforcement. There is much space for informal decision making in the latter. Every formal system involves a certain degree of informality and requires some extent of improvisation and bottom-up tinkering to make it work in the first place (Graham and Thrift 2007).

## Formalization and Language: Legality and Monitoring

More than a question of organizational arrangements, formalization manifests on a more subtle level of terminologies and legal definitions. The boundary between formal and informal is drawn by language that explicates certain things and keeps other things vague or unnamed. It starts with the legal definitions of "waste" and its ownership, and its consequences for scavenging are not limited to developing countries.

Depending on jurisdiction and situation, waste may be considered to belong to its previous holder, be part of the public domain, or be owned by the city or its service provider. According to a U.S. Supreme Court ruling in 1988, trash left out on a public street can be interpreted as abandoned by its previous owner and part of the public domain. Municipal ordinances can be more specific. Some transfer the ownership of waste to the service provider at the moment when it is collected, others at the time when an object is placed in the waste bin. In the first case, scavenging from containers is legal, in the second instance it is not. Here, it is worth focusing on the second point in the previously mentioned European definition of "waste" as "any substance or object (1) which the holder discards or (2) intends or (3) is required to discard" (European Commission 2008). In some European municipalities, the intention to discard suffices to transfer ownership to the city or service provider. In Vienna, leaving an object on the sidewalk for the city to pick up, for example, is considered an expression of this intent. Taking it from the street becomes theft, therefore rendering illegal the activities of Eastern European scavengers specializing in bulky waste.

Beyond waste ownership, many other factors influence the legality of waste picking, including the legal right to operate a business entity along with an organization's form and compliance with health and safety regulation as well as taxation. Legalizing waste picking therefore requires legislative action that recognizes and defines waste picking as a professional activity.

## Formalization through Monitoring Systems

A second criterion for distinguishing formal from informal systems is the presence of monitoring. Formalization means establishing infrastructures that capture and represent processes under defined conditions, thereby giving structure to informal practices. Monitoring encompasses data collection, storage, and analysis, as well as protocols for responses to the measured results.

As an information-gathering practice, monitoring is different from surveillance or evaluation. Management theory defines monitoring as the organized and repetitive measurement of specified parameters over an extended period (Vos, Meelis, and Ter Keurs 2000). Unlike system evaluation, monitoring is constantly ongoing. And unlike surveillance, monitoring always serves a narrow goal and implies a standard against which an activity is measured (Hellawell 1991). Management theorist William Deming noted the need to explicate and observe goals in organizational systems: "What is a system? A system is a network of interdependent components that work together to try to accomplish the aim of the system. A system must have an aim. Without an aim, there is no system. The aim of the system must be clear to everyone in the system. The aim must include plans for the future. The aim is a value judgment" (1993, 50).

Waste monitoring infrastructures may serve to improve efficiency and sustainability, verify contractual obligations, measure service quality, prevent accidents, identify causes of pollution, or shed light on consumption behavior. Most of these goals are not directly observable in individual actions, they are constructs that have to be operationalized from a range of measurable variables. Monitoring requires establishing a domain-specific language to describe and classify all processes of interest while keeping aspects outside of its scope undefined. The language for describing waste management processes is politically fraught and has direct implications on the physical reality of waste systems.

Early U.S. environmental regulations reserved the first several years of their implementation for collecting data about regulated substances, processes, and facilities. Similarly, the first stages of Brazil's NPSW were dedicated to collecting data for diagnosing the current system, registering hazardous waste operators, establishing technical standards and regional solid waste plans, and negotiating specific responsibilities in sectoral agreements. In the Brazilian model of shared responsibility, all actors are involved in monitoring and collecting data about the system. This includes civil society, as acknowledged in provisions for social control that mandate public access to information and participation in solid-waste policy making.

The NPSW also strives to make the value chain legible on the local scale. Cooperatives and associations are requested to submit documentation about materials collected and sold. The governments are expected to help cooperatives establish and formalize relationships with companies and institutions for technical assistance, education, or service contracts. Several projects that concern the professionalization of waste picker cooperatives focus on information management and accounting practices. The software CataFácil, developed by the Avina Foundation and the Federal University of Minas Gerais, supports mass balances, keeps track of work hours and finances, and can be used to generate various reports (Bazo Soluções 2015). In partnership with the Dutch government, CEMPRE, a Colombian NGO focused on recycling, develops benchmarking and monitoring tools for the value chain using interfaces that address the needs of both formal and informal actors.[1]

Although waste monitoring mandated by national laws implies centralized structures, it does not exclusively mandate a view from above. Somewhat paradoxically, practices of monitoring are not always perfectly formalized and stable, and they often involve informal aspects. Studying the monitoring infrastructures of international development projects, Casper Jensen and Britt Winthereik observe: "Monitoring moves around, and what monitoring means, what it entails, whom it involves, and how it is done all move in the same process. As this happens, initial (design) intentions and (political) ambitions for accountability and transparency are undone and redone" (Jensen and Winthereik 2013).

### Extended Producer Responsibility and Informal Recycling

A circular economy where all waste materials are folded back into production must go beyond the scope of waste management and recycling to consider all aspects of a product's life cycle, including production, consumption, and end of life. This requires extensive infrastructures of data collection.

Although extended producer responsibility (EPR) is a relatively recent concern in emerging economies such as Brazil, it is nevertheless a pressing one, considering the number of PET bottles and packaging supplies brought into circulation by international companies and the absence of a comprehensive system to recover and return these materials.

EPR policies are based on the *polluter pays* principle, which makes manufacturers responsible, partially or entirely, for bearing the cost of a product's end-of-life treatment. The policy is intended to create an incentive to design products and packaging that are environmentally benign (OECD 2001, 9).

A widespread example of EPR known as "bottle bills" instituted mandatory deposits for beverage containers to help ensure that containers find their way back to stores. U.S. states with bottle bills have substantially higher recycling rates, not least because the deposits provide income for waste pickers. But the bottle bill model does not easily extend to all material categories. More complex EPR implementations include the European Union's directive for Waste Electrical and Electronic Equipment (WEEE) and the German dual system for recycling, which has been in effect since 1991. Popularly known as the "Grüner Punkt" (Green Dot) program, the dual system was established by manufacturers and retailers in response to a federal law that required retailers to take back packaging waste. Attempts to introduce similar policies in the United States date back to the 1970s, but they have failed due to industry resistance (Marques and Cruz 2015, 18; MacBride 2012, 58; Connett and Spiegelman 2013, 248).

Like the landfill closures and sanitary disposal programs discussed earlier, EPR presents a dilemma for informal recycling. A rigorous implementation comparable to the Green Dot program collides with the interests and practices of the informal collectors who compete with the service companies for the same materials (Scheinberg et al. 2016). An EPR approach that does not displace waste pickers requires an adequate model in which both informal and formal actors can coexist, and such a collaboration requires information exchange between the actors.

Making informed decisions about eco-friendly package design requires life-cycle assessment (LCA) data models that cover the entire product life cycle, its origins and the energy footprints of its production and transportation. The same data are necessary for determining appropriate postconsumer treatment options. Acquiring such data is complicated not only by proprietary and opaque manufacturing processes, but also by the fact that many items are sold, resold, and modified by their owners several times. As a result, reliable and trustworthy LCA models for e-waste beyond the most general categories are still lacking (Offenhuber, Wolf, and Ratti 2013). An EPR approach that integrates informal recycling requires detailed information about the impacts of all processes in the value chain. As we will see in the following chapter, this is especially challenging in the case of Brazil, where the responsibilities for collecting such data are shared among multiple parties.

## The Politics of Extended Producer Responsibility

Compared to pure EPR implementations that hold manufacturers responsible for the recycling of their products and packaging materials, the Brazilian legislation could be more appropriately described as "extended product responsibility," since it divides the responsibility among producers, consumers, and the waste management sector (Migliano, Demajorovic, and

Xavier 2014). In this context, the law also makes special provisions for waste picker cooperatives, requiring companies and public institutions to collaborate and use their services. This arrangement implies a complex system of responsibilities that have to be negotiated separately through sectoral agreements.

The interpretation of shared responsibility is a political question that reveals itself in the minute details that regulate information collection and exchange. This includes issues such as how much proprietary information manufacturers must share with the waste management sector and the public. It also includes questions about what can be labeled as "recyclable" on packaging. Should this include items that are theoretically recyclable or only those that are reliably processed by recycling centers?

The New Solid Waste Policy of Brazil explicitly requires producers to "disseminate information concerning the ways of avoiding, recycling and eliminating the solid wastes associated with their respective products" (Ayoub e Silva, Leitão, and Lemos 2014, 195). This involves appropriate labeling and access to manufacturing information regardless of corporate secrecy. In the first sectoral agreement on the plastic waste stream, however, the binding labeling requirements for producers did not meet aspirations (ibid., 197). This example illustrates the limits of the shared responsibility model, which requires a complex set of agreements with powerful stakeholders who have an advantage in negotiations.

### Information Flows in Forward and Reverse Logistics Systems

In all forms of EPR, monitoring infrastructures are necessary for evaluating environmental impacts of processes. Crossing the information gap between upstream production and downstream end-of-life treatment involves political, technical, and logistical questions.

Implementing a reverse logistics system such as the one envisioned by NPSW may appear to be a straightforward application of supply-chain management, but the flow of information is quite different despite many similarities. Supply chains have comprehensive information management with well-defined interfaces for data exchange as needed by producers and retailers. Reverse logistic systems, in contrast, start where product consumption ends. As environmental scientist Marcello Veiga explains:

Many particularities of reverse logistics have been continuously ignored. Forward logistics is an active process, where firms plan, produce, and supply distributors with products based upon forecasts. Reverse logistics is a reactive process with more unpredictable factors, which is usually initiated by end-user. The main trigger for reverse logistics process is the end-user, not the manufacturers themselves. (Veiga 2013, 653)

This has the consequence that little data are readily available upon collection. Products that have passed through various stages of use and reuse pose problems for identifying manufacturers and collecting information about material composition. Consequently, monitoring practices that focus on efficiency, transparency, and contractual compliance are not sufficient. Any successful reverse logistics system depends on the voluntary involvement of the individual, whether scavenger or resident, and needs to consider questions of motivation, social practice, and culture.

**Local Legibility: Context-Oriented Data Initiatives and Projects**

Understanding a system such as the value chain, its actors, and their links to the larger waste system requires a different kind of legibility, one that recognizes material collection and valorization as local activities that depend on specific local conditions.

While some materials can be collected from the street, the recovery of others such as electronic devices requires the engagement of local residents to capture information that links these devices back to their manufacturers. Waste pickers respond to local conditions in the form of prices paid by junk shops, the supply of materials, and the available options for processing.

Waste picker cooperatives require information for managing their operations, but data collection has to take their practices and local context into consideration since they remain the most vulnerable party in the value chain. The value chain itself consists of local processes with intricate dependencies that are easily destabilized by changing markets and local actors. Monitoring through the local community, which is specified as "social control" in NPSW and delegated to municipal governments, again requires the consideration of local specifics.

The regulatory instruments of national programs like RCRA can be instrumental in evaluating the overall waste system and its impacts. They benefit from harmonized definitions of waste across localities, technical standards, and common data formats for data exchange. Their systemic scale is not sufficient at the local level of the value chain and its actors, which requires approaches that have their own modes of legibility within their specific contexts.

The importance of the local context can be demonstrated with the example of Brazilian recycling cooperatives and associations I describe in the next chapter. Currently, less than 10 percent of Brazil's waste pickers are organized in such groups, although NPSW encourages more of it. Despite the advantages of collective action, each cooperative remains a precarious achievement that depends on specific conditions such as a suitable facility site, a sufficient source of waste material, the engagement of individual waste pickers, or the support of local politicians. No formula for success is easily transferred because they all have their own stories, local contingencies, and business strategies. From an information standpoint, each cooperative requires its own window into its operating environment to evaluate the costs of collection and the value of their service.

## Local Data

Media scholar Yanni Loukissas argues that all data are local because they require a local reading that takes the context of their generation into account. What may appear as a large archive of uniformly structured data reveals itself under scrutiny as a heterogeneous collection whose artifacts bear traces of the conditions of their generation. Classifications are always designed with a specific problem in mind, and typos, duplicates, and other artifacts reveal the methods of data collection and entry. Local data are therefore not a different kind of data source, but a different lens (Loukissas 2017).

Many recent experimental projects focusing on waste issues are local in their scope. Some of them might seem whimsical, such as the idea to locate open dumpsites by outfitting vultures with camera backpacks,[2] or to identify litter by using image recognition algorithms on CCTV footage.[3] Crowdsourcing initiatives that use technologies that appear to be generic, scalable, and location independent still depend on the engagement of a local community.[4] The importance of the local context, however, does not imply that projects are necessarily limited to it. Projects that involve collecting recyclable materials with a fleet of digitally coordinated bicycles exist in Nigeria,[5] India,[6] and Massachusetts.[7] Open source and civic technology initiatives are anchored in a local context while simultaneously engaged in a constant exchange with similar initiatives around the world.

## How Can Formalization Support Recycling Cooperatives?

The circular economy envisioned in EPR schemes creates and reinforces an economy of information. When provenance data is necessary to link a product to its responsible party, items of electronic waste with well-documented sources yield higher prices than undocumented scraps because they can be sold and processed in a more formal setting. This affects not only recycling cooperative profits but also the selection of potential customers. In the Brazilian shared responsibility model, waste pickers may not have the same rigid reporting duties compared to formal companies, but information remains critical for their partnerships with companies. Operational data are also necessary for cooperatives to demonstrate the value of their service in their political negotiations with cities, helping them to shift the theme of waste picker inclusion from charity (under the motto "anything helps") to an adequate compensation for services, which can be in many ways superior to professional recycling.

The next chapter addresses the question of how waste picker cooperatives can design information infrastructures that are tailored to these specific needs. Cooperatives have to make informed decisions about where and what they collect. An experienced individual collector with a handcart may be able to make a living by picking the right material in the right place and selling it to the right buyer at the right time, but in a larger cooperative with machines, trucks, and a big collection area, the best strategies of valorization are no longer obvious. Most cooperatives have a good sense of their costs and opportunities, but they also know which information they lack. In the following chapter, I describe a participatory design approach used with different Brazilian recycling cooperatives to investigate information management and to explore opportunities for using ubiquitous communication technologies to implement data collection infrastructures that respond to the cooperatives' needs.

# 4 Tacit Arrangements: Reading Presence and Practices

The size of the informal economy in Brazil is significant: estimates range from 40 percent to over 55 percent of the national economy (Budlender 2011; Henley, Arabsheibani, and Carneiro 2009). Around five hundred thousand waste pickers collect sellable material from streets, buildings, and dumpsites (Fergutz, Dias, and Mitlin 2011), accounting for over 90 percent of the material recycled in Brazil (Medina 2007a, 70). The pickers started unionizing and forming cooperatives during the 1980s. Today, about 10 percent of the country's pickers are organized in over five hundred cooperatives with a total number of sixty thousand members. Known as *Catadores de Lixo*, the organized waste pickers have formed a political arm to support their cause in the public arena (Medina 2010).

## Steps toward Formalizing Waste Picking in Brazil

As in most countries, solid waste management in Brazil is primarily the responsibility of municipalities. The first municipal ordinances recognizing and integrating informal waste management were instigated during the 1990s in the cities Belo Horizonte, Porto Alegre, and Diadema (Medina 2008). During the early 2000s, many states closed open dumpsites, which were deemed hazardous but provided livelihoods for thousands of waste pickers who lived and collected material on the sites. To mitigate the social impact of displacing waste pickers, inclusive waste policies have been formulated by many states. In 2002, the Brazilian Ministry of Labor and Employment incorporated the profession of the *Catador de Material Reciclável* into the Brazilian Classification of Occupations (CBO), including a taxonomy of fifty different tasks falling to this profession (Dias 2009, 2–3).

With waste production increasing over 90 percent during the past twenty years and landfill space diminishing due to the closure of inadequate open dumpsites (Veiga 2013), waste management in Brazil has become a national priority. On August 2, 2010, the National Congress of Brazil passed the National Policy on Solid Waste (NPSW),[1] a law that had been in the making for twenty years (Brazil 2010). As the first national waste legislation and the latest step in a long process of formalization, it brought significant change to waste management and recycling. Written with the participation of the National Movement of Waste Pickers,[2] it recognized the work of waste pickers and integrated them into the waste and recycling system, granting cooperatives access to private and public contracts and requiring institutions and companies to use their services (Ministerio do Meio Ambiente 2010).

The NPSW is somewhat vague about its expectations from cooperatives, starting with the fact that it does not define specifically what a "recycling cooperative" is. According to federal law, a "cooperative" must have a minimum of ten members and meet specific accounting and procedural requirements—both requirements can be a challenge for smaller groups (Santos et al. 2014, 215). Currently, only a few recycling cooperatives are able to satisfy these requirements, often by relying on family members to form the basis of the cooperative.

Formal recognition and access to service contracts alone, however, are not enough to improve the situation of cooperatives. Additional support is necessary. Recognition can be used as a pretext for introducing additional bureaucratic hurdles; recent efforts by political opponents to tighten the regulation of recycling cooperatives under the guise of empowerment were opposed by the National Movement and subsequently vetoed by former President Dilma Rousseff (Senado Federal 2012). Most recycling cooperatives operate without formal planning and accounting, constituting what social anthropologist Jean Lave and computer scientist Etienne Wenger describe as a community of practice, in which members learn informally from each other (Lave and Wenger 1991).

### The Data Requirements of Formalization

The tension between the opportunities and burdens of formalization is illustrated by a central provision of the NPSW that implements extended producer responsibility. Toward this end, the NPSW mandates the implementation of a reverse logistics system for organizing the recovery of electronic and hazardous wastes. As discussed in the preceding chapter, reverse logistics requires information infrastructures for collecting product lifecycle data, including information to track products through the removal chain—no easy task, given how many orphaned and reappropriated products pass in and out of local markets (Scartezini 2013; Consonni 2013).

As a result of the reverse logistics system, discarded electric and electronic appliances with documented sources yield higher prices for waste pickers. Although collecting provenance data is primarily a task of the waste management and recycling industry, it affects cooperatives selling material to these companies. Currently, there is little reason for cooperatives to pursue this recycling market because it is relatively small and the prices for e-waste disposal are low. But this might soon change: Brazil has become the fifth-largest electronic market after China, the United States, Japan, and Russia (Streicher-Porte 2009).

The information requirement of the NPSW and its effect on material prices and available buyers illustrates the importance of data, even for small, informal enterprises. Policy measures such as the SIMPLES and SIMEI programs, intended to simplify data reporting and taxation for micro-firms (Brazil 2006), were largely considered effective at increasing revenue and profits of micro-firms (Fajnzylber, Maloney, and Montes-Rojas 2009). Nevertheless, accounting and data collection remain challenges in an environment where workers lack basic education. Educational projects aiming to help waste pickers develop skills, such as logging their work hours and keeping track of the amounts of materials sold, are still at an early stage (ITCP-FGV 2012).

The unintended implications of information requirements in the NPSW highlight how difficult it is to evaluate the effectiveness of social policies designed to improve the lives of pickers. All efforts at formalization, whether inclusive or prescriptive, require legibility, the collection of information as a basis for policy decisions.

## Formalization Models for Recycling Cooperatives

Many details in the work of recycling cooperatives and associations remain unregulated, which benefits waste picker groups and encourages experimentation with different models. Waste picker organizations in Brazil show a variety of different organizational structures, form dynamic alliances in cooperative networks, and operate with different business models. Brazil's approach to formalization could be described as *market-* and *partnership-oriented*. Collectors of recyclable materials are mainly organized in two forms of nonprofit entities offered by Brazilian legislation: associations and cooperatives. Associations are typically nonprofit charitable organizations that can render services to society as well as their own members. Profits made from these services have to be reinvested in their operation. Unlike associations, cooperatives can offer for-profit services, and profits have to be distributed among their members. Compared to associations, cooperatives require more resources, complex governance structures, and reporting requirements (Gouvêa and Monguilod 2014). Beyond these legal entities, a number of organizational models have emerged; Anne Scheinberg's distinction between the service model, commodities model, and hybrid models for inclusive recycling are useful in this regard (Scheinberg 2012).

The most common arrangement for larger cooperatives is a commodities model in which waste pickers earn income by selling recyclable material. In the service model, less common in Brazil, waste pickers are paid for their work (collection, sorting, baling), but not the materials they collect. Aspects of community-based enterprises are also present to various degrees, expressed in complex partnership structures around cooperatives.

In the commodities arrangement, local or regional governments often own the recycling cooperative's facility, initiate partnerships with other institutions and companies on their behalf, and provide services such as water and electricity. Private companies enter into collection or sorting contracts with the cooperatives, often sponsoring equipment such as trucks and machinery. Business associations such as CEMPRE[3] coordinate private partnerships. NGOs and public institutions may provide legal or educational services, as well as enter into collection contracts as mandated by the NPSW.

The national government promotes partnerships between cooperatives and the private sector, connecting the cooperatives with companies and providing incentives for businesses to sponsor machines, facilities, and training programs. An example of such a partnership is the Cata Ação[4] (Collect Action) project, a training program for cooperatives that involves familiar players in the Brazilian development sector such as the Avina Foundation,[5] the Inter-American Development Bank,[6] the Coca-Cola Institute,[7] nonprofit foundations, the waste picker movement, and the national government.

**Economic Situation of Cooperatives**   Judged by their equipment, size, and organizational structures, the advanced cooperatives seem indistinguishable from small, formal enterprises. Despite these similarities, cooperatives work under very different conditions and structural constraints compared to formal recycling companies. When cooperatives are not paid for their services, members are expected to make a living by selling the materials they collect, a livelihood that fluctuates with the market. As of 2009, members of the COOPAMARE cooperative in São Paulo earned twice the country's minimum wage, although that is not a standard situation (WIEGO 2013). Many cooperative members live in poverty, lacking basic education and earning less than minimum wage. The cooperatives are well aware of these nuances and contradictions.

Support for cooperatives from local and regional governments comes in the form of grants for education, business management, or housing. Healthcare is provided by a public system available to all Brazilian citizens. Municipalities rarely pay the cooperatives for services such as curbside collection, acting instead as supervisors and facilitators for private partnerships. They typically view public contracts for cooperatives as a form of social welfare. Rather than being seen as beneficiaries of a cooperative's service, waste generators are often framed as "material donors," a term that can be found in official documents (Diário de Pernambuco 2013).

**Professional Aspirations**  Early waste picker collectives deliberately focused on commodities and did not actively pursue involvement in service provision. As cooperatives took on larger projects such as collection during the events of the 2014 FIFA World Cup or the 2016 Olympic Games, many of them no longer wanted to be framed as beneficiaries of donated materials but as waste management professionals who had to be paid commensurately with their services. A frequent complaint by waste pickers is that cities pay large sums for service contracts with waste management corporations, but expect groups to subsist on the revenue of their collected material. The commodities model, however, is unlikely to disappear in the near future. The Brazilian waste management sector is becoming increasingly data driven under NPSW. In addition, material prices are no longer only determined by volume, purity, and commodity value, but increasingly also by information.

**The Role of Digital Technology in Formalization**  In addressing questions of how information management can assist cooperatives with professionalization, digital technology plays a crucial, if vexing, role. Information and communication technologies can informalize and formalize at the same time. By simplifying and extending the reach of communication, location-based digital technology can make the coordination of large groups in geographic space more informal. At the same time, all interactions can be recorded, making them more formal. The question stands whether digital technology can contribute to making an informal sector more formal without crystallizing its work into rigid structures, a question we investigated with the Forage Tracker project.[8]

## The Forage Tracker Experiment

The phenomenon of waste picking has been studied by others from the perspective of economic development by measuring the quantitative impact of scavenging on the larger economy. It has also been approached from an ethnographic perspective that describes the livelihood, conditions, and survival strategies of individual waste pickers. Between this macro perspective of economic impact and micro perspective of the individual situations, a middle ground detailing the geographies of scavenging remains largely unexplored. What is missing is a picture of the spatial dimension of scavenging: the reach and coverage of intricate networks of scavengers, scrap dealers, and sellers, all the way to the warehouses of companies that buy recycled materials.

The Forager Tracker experiment attempted to make informal waste systems in Brazil legible by mapping their intricate network of spatial and social practices. The name refers to *foraging theory*, a mathematical model for optimal foraging and scavenging among hunter-gatherer societies, which we naively assumed to be useful for understanding the spatial logic of scavenging in an urban environment (Charnov 1976). Of course, the various dependencies of waste picking from other actors and institutions resisted any representation through a reductive mathematical model, and foraging theory quickly moved into the background.

In setting up the project, we assumed that the informal practices of collecting and reselling that cooperatives perform are different from those of traditional waste management companies that sell their services under contractually defined conditions. Informal waste pickers collect in an opportunistic manner, specializing in specific materials and areas. They often focus on aluminum cans or corrugated cardboard for their low weight and high value. The project goal was to explicate what drives the decisions to focus on a particular area, material, and collection process. We hoped that mapping tools for data collection and spatial coordination would allow cooperatives to improve their own collection activities, document their institutional knowledge, and use the collected information to increase their public visibility. Given these aspirations, the project addressed the question of to what extent location-based technologies can support the professionalization of cooperatives without ossifying their practices.

At first, it seems absurd to look for a technological fix when members of cooperatives face more pressing social, political, and economic issues on a daily basis. Our informants in the cooperatives were certainly highly aware of all these matters and the limitations they pose for their work. However, precisely because of these narrow constraints, the cooperatives are very attentive to the intricate details of their daily operations and their effects. As previously mentioned, questions of data collection, whether for the cooperative itself or for others, are gaining importance for many cooperatives. Most cooperative members own cell phones, many of them simple smartphones. A recent study in India, South Africa, and Peru found that many waste pickers and other members of the working poor have adapted cell phones to organize their daily work (Chen 2016). Issues of legibility surfaced for the cooperatives in data exchange with clients, training for members, measurement of collection costs, and interactions with governments and residents.

## Information Collection Practices in Cooperatives and Associations

The recycling cooperatives we interviewed collect information about their activities for various purposes and report it in different formats. Cooperatives typically send monthly reports to the respective municipality detailing the amount of material processed. This can happen in a more or less formal way. Large cooperatives have legal representation and accountants; small groups submit handwritten reports by themselves. This includes associations, which report about processed material to the city, although they are not allowed to render for-profit services; write invoices for material sold; and sell exclusively to intermediaries. Several workers saw reporting as a mere ritual of little significance; one association did not keep copies of the forms submitted to the city each month. Interestingly, a respondent in the municipality of Recife expressed a similar sentiment, caring less about what is reported than about the fact that reporting takes place at all.

For two cooperatives in the Recife area, reporting is a source of pride. The cooperatives participated in a training program on bookkeeping/accounting and still honor this practice. Books are not hidden in an office drawer, but prominently displayed in the common room—both as a tool for internal accountability and a signal of their new formal status as a cooperative.

Reporting has a different status for cooperatives when it does not involve the city, but a company. Data exchanges with businesses help cooperatives secure service contracts or sell material that they would have to sell otherwise at a lower price to intermediaries. Recognizing the value of long-term contracts and partnerships, cooperatives were keen to find ways to improve their data collection and reporting methods. Two of the cooperatives have started to use software developed by the Avina Foundation and the Federal University of Minas Gerais (Bazo Soluções 2015). The software CataFácil supports mass balances, keeping track of work hours and finances, and can be used to generate various reports.

Practices of internal information management in the cooperatives are hybrid and involve a variety of different formats: calendars for documenting collection routes and work hours or, perhaps, notes on bulletin boards specifying the latest prices for materials. In some cooperatives, the respective municipality and accounting, oversight, and route planning are in the hands of public officials, who often also provide logistic and legal support to the cooperative. Transportation costs are rarely a concern for cooperatives that operate trucks, because maintenance costs and fuel are usually covered by the city or through a private sponsorship. The profits from selling material would not be sufficient for both maintaining a vehicle and covering the wages of the cooperative members.

**Figure 4.1**
Hand-written material price list at the Associação de Catadores O Verde é a Nossa Vida facility. Recife, June 2013, photo by the author.

**Participatory Design Workshops on Information Management**
In the course of our fieldwork, which spanned three years, we examined the existing information management practices in the cooperatives as well as the challenges they faced with coordinating information. The addendum to part II details the histories and approaches to information management in different cooperatives and associations. In this chapter, I will focus on a series of participatory design workshops with the cooperatives, in which we aimed to design, prototype, and evaluate a platform for community-based recycling through which clients could submit information about available material and cooperatives could plan their collection more efficiently. In the design process, three goals emerged:

1. Document collection routes and areas to look at potentials for coordination within the cooperative.
2. Investigate ways to simplify the communication between the cooperative and its clients.
3. Explore tracking approaches for electronic waste.

Practices of participatory design (PD) and critical making guided our approach. PD is a methodology for involving users in the design of sociotechnical systems (Kensing and Blomberg 1998). Critical making refers to a reflective practice of prototyping and developing with the goal of producing artifacts that stimulate debate (Ratto 2014). Both practices are outcome- and process-oriented. They build artifacts while learning about the issues relevant for users. While avoiding an unreflective positive approach to technology, both practices also refrain from acting like disengaged, critical commentators.

While it would be highly desirable if a design workshop would result in a tool of practical use for the cooperative, it would be naïve to expect any single solution would solve all problems of information management. However, in the process of making and reflecting on prototypes through multiple iterations, we discovered and examined issues that otherwise would not have been obvious. Building and discussing prototypes also helped both us and the cooperatives to calibrate expectations toward the use of technology. By examining the hypothetical scenario of the technology-mediated collection that connects collectors and material sources, we also were able to better understand nontechnological issues.

In the earlier Trash Track project, we used GPS tracking as a mode of passive observation, an approach that seemed inappropriate in this context. In working with the cooperatives, location sensing became a mode of active documentation in which the collector retained agency and control. To avoid a scenario of real-time surveillance, the sensors carried by individual collectors did not report back over the wireless network. Furthermore, the recorded traces were interpreted based on the explanations of the collectors. The GPS traces served as prompts for detailed interviews and mapping sessions with collectors aimed at better understanding the spatial logic of collection. By contextualizing the recorded spatial traces with the collector's explanation for choosing a particular route at a given time, we gained glimpses into how collectors read the city and into the geographies of waste collection that result from factors such as traffic, terrain, distribution of material, distance from the cooperative, and landmarks such as regular clients and junk dealers.

Besides the GPS loggers carried by manual collectors, we also used smartphone applications for coordinating collection trucks. My colleague David Lee and I spent several days working as collectors ourselves, filling the cooperative truck with material from different collection points.

Although using a truck that can hold six tons of commingled materials may seem vastly superior to manual collection, we learned from these experiences that truck collection faces several constraints, including traffic, lack of parking, and a night-time truck ban in central São Paulo. Collection from a remotely located client often yielded an insufficient amount of material that did not justify the trip; delays, traffic jams, and occasional miscoordination between the collector and the client complicated routing efforts.

Understanding how waste picking operates requires looking at multiple organizations because each cooperative has a unique spatial collection pattern. Smaller groups often collect material opportunistically from the street. Larger groups can enter into contracts to collect from the curbside using trucks or handcarts. Some cooperatives predominantly collect from drop-off points or clients such as supermarkets and companies rather than off the street.[9] Other groups do not collect at all, sorting and baling the material delivered by the municipality or businesses instead. Sources of income vary for each cooperative, which may sell sorted material to the municipality, recycling companies, or intermediaries such as junk shops.

After preliminary surveys of different cooperatives over the course of a year, we selected two cooperatives, COOPAMARE in São Paulo and Pró-Recife in Pernambuco, for in-depth participatory design workshops. Both cooperatives were relatively developed organizations that used computers for managing their data, planning routes, and submitting reports. We later tested the smartphone-based collection route mapping prototypes designed in these workshops with other cooperatives and associations in their respective cities. The issues identified during the workshops and site visits highlighted areas where data coordination can help cooperatives work more efficiently, earn more money, and cope with data collection needs.

## São Paulo

With a fluctuating number of around eighty members, COOPAMARE is relatively large. Founded in 1989, the cooperative has a long history and played a central role in the Brazilian waste picker movement. The cooperative is located in the wealthy and central district of Pinheiros—an area that offers plenty of collectable material, but is unaffordable for many of the workers, who live on the outskirts of the city.

The cooperative follows a model of selective collection, receiving material from selected clients including supermarkets, condo buildings, and public institutions. Following a commodities model, the cooperative sells material directly to recycling companies. As detailed in the addendum to part II, this model is one of several approaches to collection. In comparison, the cooperative CRUMA in Poá, a city close to São Paulo, collects material from a designated area rather than selective clients, using a system of manual carts and a supporting truck owned by the cooperative. CRUMA receives service contracts from other waste management companies. Finally, COOP-RECICLÁVEL in the city of Guarulhos, close to São Paulo's international airport, represents a hybrid model, receiving and sorting material collected by a municipal truck. But in all cases, the sale of sorted material remains a central source of income, and cost and efficiency of collection, a decisive factor.

**Route Documentation**  In our workshop with COOPAMARE, we focused on documenting collection routes and facilitating communication between the cooperative and its clients. We gave all collectors small GPS loggers that recorded manual collection routes and truck trajectories for more than a week. Each day after collection, we sat down with the collector and went through the recorded trace data to learn why specific routes were taken or certain areas visited. For these interviews, we developed an application that would show the trace on a digital map and allow us to annotate a specific part of the trace with the recorded narrative of the collector. Notably, the collectors had no difficulty reading a "naked" trace and identifying locations even without an underlying map. Their intimate knowledge of space allowed them to reconcile an abstract line representing their path in bird's-eye view with the experience of daily collection.

**Figure 4.2**
Collected GPS traces of COOPAMARE collection routes. In blue, the collection truck routes, in orange, the manual cart collection routes. São Paulo, November 2011. DigitalGlobe, reproduced under Google Maps fair-use policy; courtesy MIT Senseable City Lab 2011.

**Collection Strategies**    One collector, going by the name of Chico, repeated the same seven-kilometer route with little variation up to three times a day, returning with a handcart filled with flattened, compressed cardboard that weighed several hundred kilograms. The lightweight cart, fashioned from reclaimed wood and corrugated plastic, was his own design specially tailored for cardboard collection and featured a system of bungee cords for securing material on the cart. By specializing exclusively on one material collected from a single route, Chico was able to earn twice as much than by taking part in truck collection and splitting revenue with everyone involved.

Chico explained that he settled on his route, which was within walking distance of the cooperative, because it was commercial and had many shops that produced large quantities of cardboard. He also had a number of fixed clients he visited, and he avoided purely residential areas due to the smaller amount of material, the hilly terrain (although he used to collect there when he was younger), and streets with heavy traffic.

Elisanete, a second collector, had a smaller collection area; she gathered cardboard, metal, and PET mostly around a large street market in Perdizes. A third collector specialized in different kinds of metal that he collected in a radius of one to two kilometers around the cooperative. Successful foraging for all of these collectors depended on personal connections with clients and the careful calibration of a range of spatial and economic factors.

Although most activities in the cooperative are coordinated through the president, individual members kept a degree of autonomy to follow their own procedures. Manual collection with handcarts still exists in COOPAMARE mainly for this reason. In other cooperatives, handcart collection plays an even larger role.

**Truck Collection**    Truck collection was guided by a different logic. COOPAMARE has two trucks, but at the time of our fieldwork, only one person had a driver's license—former cooperative members who completed the course accepted different jobs after they had passed the exam. Laerte, the driver who had stayed loyal to the cooperative for over twenty years, kept a detailed handwritten list of clients and collection points that he would visit each day. We started by geocoding this list, as well as the collection frequency at these locations, and joined in as collectors on a number of collection trips. The resulting data set was merged with the traces recorded during truck collection and made available to the cooperative using a web application based on the Ushahidi platform.

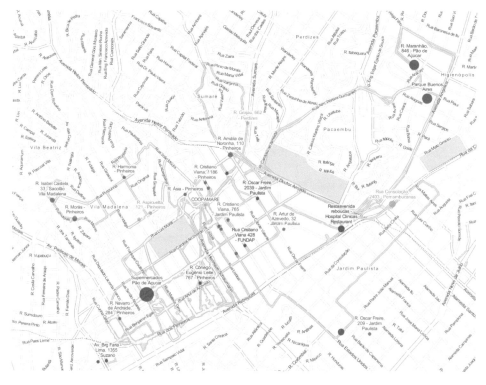

**Figure 4.3**
Visited collection points by COOPAMARE during one week of collection. Yellow traces correspond to manual collection, blue traces to truck collection. Dots represent different clients, their size indicating the frequency of visits. Courtesy MIT Senseable City Lab 2011.

Collection by hand and by truck have different advantages and fulfill different purposes. The traces rarely overlapped, illustrating how well the manual collectors and the truck complement each other, each focusing on particular strengths of the respective modes of collection. The truck driver was interested in the possibility of notifying manual pickers about material he saw on his route. In addition, he was the only collector who documented his route using a handwritten journal. He saw GPS mapping as a valuable way to share information across the cooperative.

**Real-Time Scheduling Apps**   The workshop explored two additional ideas. The cooperative's president at the time of the project, Maria Dulcinéia Silva Santos, imagined a tool that tracked the truck's location in real time, allowing her to redirect the truck if a nearby client called about a job. This was not immediately possible to realize, since it required an expensive data plan and a more stable Internet connection than was available in the cooperative, but the idea became a central aspect of the subsequent prototypes. The second idea was to provide an app through which local residents and businesses could report available material to the cooperative, allowing collectors to plan their trips more effectively.

The latter idea was implemented, but ultimately abandoned. Nevertheless, it highlighted an interesting issue: the blurry line between submitting information about a resource available and requesting a service. The cooperative feared that the convenience of the application would raise the expectations on the side of the residents that a collector would immediately show up and take the material—an expectation that the cooperative was not able and did not want to fulfill. The result would likely be a loss of trust in the work of the co-op. In order to earn their living from selling materials, the cooperative members had to be highly selective about what and where they collect.

The issue of trust was at the heart of a second concern regarding the implementation of a mediating technology. Up to 25 percent of the material processed comes from residents who drop off material at the cooperative personally. The "gatekeeper" at the cooperative who receives these visitors and accepts their material fulfills an important role as the main contact between residents and the cooperative. Residents may have questions how they should separate material for the cooperative, and COOPAMARE puts considerable effort into educating residents about how to prepare recyclables. The cooperative considered these moments of personal interaction crucial for building trust with the community, and mediating this interaction through digital technology would reduce these valuable face-to-face encounters (Offenhuber and Lee 2012).

In considering the relationship between the cooperative and the local community, one has to keep in mind that the cooperative is located in a wealthy central neighborhood, and most *cooperados* live far outside of it. During the workshops, we also mapped the daily commutes of workers, which frequently exceeded two hours in a single direction. It is therefore important to examine practices of informal recycling in relation to other social issues, such as housing and education. Many large cooperatives address these matters in some way. Members of COOPAMARE were involved in the development of a housing project in downtown São Paulo, and CRUMA operates a school for computer literacy in connection with its e-waste recycling and refurbishing program.

## Recife

The recycling cooperatives around Recife have a shorter history, and the field is less settled than in São Paulo. The state of Pernambuco implemented its first solid waste policy in 2003, which laid the foundation for the current implementation plan required by the 2010 National Policy. The state policy offers financial support to help cities with adopting inclusive waste policies (Agência Estadual de Meio Ambiente 2012, 196). Currently, the state government is collecting data on the characteristics and qualifications of its waste picker workforce. It plans to set up a business incubator, an institution to support the development and training of new cooperatives.

The different administrative responsibilities for waste management present obstacles for harmonizing data across the state. In rural areas, waste issues are frequently administered by departments of agriculture, while in urban areas, they fall under departments of environmental protection. Efforts to integrate the diverse waste policies were galvanized by the imminent closure of the region's main landfill in Muribeca, Jaboatão, which until 2010 served fourteen different municipalities and was occupied by more than three thousand waste pickers. A series of protests in 2004 launched statewide efforts and social initiatives to find other occupations for the displaced waste pickers (Ministério Público de Pernambuco 2009).

**Recife Cooperatives and Associations**   In the second series of workshops in Recife, we modified the technologies based on what we had learned in São Paulo. We developed our tools primarily with the cooperative Pró-Recife, but also tested them with a number of other cooperatives that are described in detail in the addendum to part II:

• Pró-Recife operates a selective collection service with individual clients, including public institutions and companies. It operates two trucks, collecting material from all over Recife six days of the week. The cooperative sells directly to the recycling industry.

• COOREPLAST and COOCARES in the town Abreu e Lima have separate facilities but operate under joint management. Together, they operate two trucks and take weekly turns in the collection. COOREPLAST and COOCARES sell to both intermediaries and industry.

• ARO in Olinda collects recyclables and waste from the curbside and permanent collection points using a truck and carts. The group represents a service model in which the city takes care of logistics, organization, and oversight, and buys sorted material at fixed prices.

• O Verde é a Nossa Vida is a small association in Recife, selling material to intermediaries. They receive material delivered by companies and collect using manual carts.

**Fleet Management** Pró-Recife, like COOPAMARE previously, expressed an interest in real-time fleet management. Because of the availability of cheap smartphones and prepaid data plans, we chose to develop an app that ran on older, cheaper Android hardware like the phones already used by some of the Pró-Recife collectors. The system was designed to document the collection routes and pickup locations along with information about collected materials. The software prototype continuously tracked the location of the phone unless the collector disabled the feature. It also allowed the collector to catalogue materials by taking a picture of the collected items and optionally tagging objects of interest such as e-waste with a barcode sticker.

Pró-Recife saw the value of documenting the collection process as a way to improve internal management and raise external visibility. By tracking collection activities, they would gain a better method for estimating the cost of collecting from a specified location compared to the value of the goods recuperated. The cooperative was highly interested in tracking their truck in real time and being able to contact the collection team to improve coordination with clients at pick-up locations.

Testing our prototype in the different cooperatives revealed a number of issues. Collection stops are usually brief to prevent a truck from blocking traffic or remaining at a loading dock too long. The size of the truck used by Pró-Recife (larger than the one at COOPAMARE) complicated the logistics. Each collection trip involves up to twenty collectors and could take an entire day. Pró-Recife schedules routes on a computer, but the actual routes often differ for several reasons. Traffic in Recife is more unpredictable than in São Paulo. Torrential rains fall almost daily from March to June, often making roads impassable due to flooding and deep potholes.

**E-waste Tracking** Tracking and documenting e-waste was an issue of particular concern for Pró-Recife. The cooperative frequently receives devices and appliances such as old monitors or TVs. Without documentation of their origins, these devices are not accepted by recycling companies. Consequently, they have to be sold to junk shops for prices that are often lower than that of PET plastic bottles.

During the workshops, we developed a simple Android smartphone app for the collection truck that would document collection and offered

**Figure 4.4**
Screenshot of Android application for recording trajectories and documenting collected material. Courtesy Julian Contreras, David Lee, MIT Senseable City Lab 2013.

the driver simple options for adding stops and estimating the amount of
material collected. In addition, individual devices could be registered and
tracked. Documenting the origin of these devices may seem conceptually
simple, but involves multiple steps that are often difficult to execute during
collection in the field.

In the experimental app, the collectors had to take a picture of the whole
item, attach one of the prepared barcode stickers, and take another picture
of the sticker to register the item's location with the barcode ID. Even if not
all steps were executed as intended or the app failed, it was often possible
to reconstruct missing information. As was not unexpected, this feature did
not survive the field test due to the multiple steps necessary, and it turned
out to be easier to catalogue the e-waste object in the cooperative after the
collection trip. A procedurally simpler, but technologically more complex
approach was tested that involved a mobile printer to print stickers with
the encoded time and GPS location of the collection point, which could be
attached to the item.

**Privacy** According to our interviews, information exchange between
informal collectors is characterized by both competition and cooperation.
Collectors typically safeguard information about their favorite collection
spots from other collectors while sharing information about other materials
and the prices offered by different intermediaries.

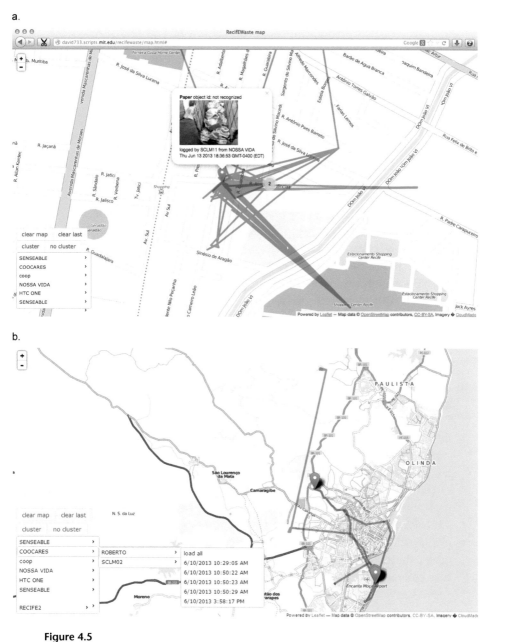

**Figure 4.5**

Online map for viewing trajectories and recorded objects. Web application: David Lee, MIT Senseable City Lab 2013.

During the Forage Tracker experiment, the participants viewed information exchange among different cooperatives as desirable and a prerequisite for setting up a network of cooperatives in Recife. Although each participant could switch off tracking using a prominent button in the app, the collectors rarely used this option, though the limited scope of the experiment did not allow us to investigate this issue in detail.

**Safety and Interface Issues**   The field test in Recife revealed a number of problems. For manual collection, safety is a challenge. In the Boa Viagem district where Pró-Recife is located, collectors did not carry a phone out of fear of robberies. For others, the multiple steps necessary to input the data, starting from switching on the phone, unlocking the screen, and launching the application, presented problems. Bright sunlight, humidity, and dirt were obstacles for operating the phones during collection.

Over the course of the workshop, we continuously simplified the interface. Since some collectors had difficulties reading, we labeled interactive elements with pictograms rather than text. We noticed that some of the illiterate collectors still owned smartphones and navigated their user interface seemingly without effort. One respondent claimed that his illiterate colleagues were more successful navigating the interface of our prototype than the more literate collectors.

Memorizing a visual interface to accomplish a specific task is one thing, but gaining a conceptual understanding of technology is quite another. The concept of real-time data was intriguing and sometimes confusing for members of the cooperative. A participant of the Recife workshop expressed surprise when a photo that was taken with an app instantaneously appeared on a screen in a different part of the room.

During the testing period of the prototype, we encountered technical problems that included unreliable localization, weak cell phone reception, a broken car charger that caused the smartphone battery to drain, a data plan that kept running out of bandwidth despite our best efforts to minimize data traffic, and the challenges of remotely troubleshooting issues as they emerged.

Pró-Recife was encouraged by the field tests nevertheless and offered ideas for developing the system. We settled on a solution that installed smartphones in the truck to record routes, enable communication with the cooperative, and record additional data about material collected at individual stops. At the cooperative's suggestion, we prototypically implemented a function to compile the recorded data into a spreadsheet to simplify interaction with material buyers.

## Data Management Challenges for Cooperatives

Our fieldwork in São Paulo and Recife was useful for investigating the problems that cooperatives face with information management, challenges that can be grouped into three main areas, described in the following sections. In most of these aspects, improved data management can help cooperatives run their operations more effectively, increasing the legibility of their activities for themselves and their partners and establishing more formality within the organizations. Data sharing within regional and national networks can help the cooperatives learn from each other. On the downside, it can expose critical knowledge and the weaknesses of their services to local governments, competitors, and clients.

### Costs of Collecting Commodities Are Hard to Determine

When the operation of a cooperative entails making its money by selling the material it collects, the question of whether a particular area or client is beneficial requires a close look at the spatial dimensions of collection. Under the commodities model, waste picking entails the economy of movement. A locale may offer plenty of material that can be valorized, but searching, collecting, sorting, and selling come at a cost. Pickers must constantly assess potential areas and decide what to collect. Their way of reading the city considers not only topography and traffic, but also factors such as the weight, bulk, and value of material, as well as the whereabouts of intermediaries and competition.

Although waste pickers know their collection areas well, maps of informal settlements often do not exist or are inaccurate. We discovered that an entire neighborhood where collection took place was missing from online maps. At one cooperative, a single person kept track of collection schedules using a paper journal. Even when schedules were planned using a computer, the cooperative leaders noticed large discrepancies between scheduled and recorded routes. Traditional contracts for servicing an area are hardly possible under these conditions.

When a cooperative of twenty or more members operates in a large collection area, the tacit knowledge of the individual picker is no longer sufficient for success, but the true costs of collection and valorization are difficult to determine without a detailed look at day-to-day practices. To understand which practices are profitable, cooperatives must be able to plan routes efficiently and determine the distance and value of reclaimed materials. In their daily work, cooperatives need these data to assess their real costs of collection.

### Collection Services Are Frequently Not Profitable

Although contracts for curbside pickup are attractive because of the long-term stability offered by a service situation rather than the fluctuating market for commodities, they can prove less valuable because they restrict the freedom to handpick the most profitable items. Service contracts are based on the expectation that all material is picked up whether or not it holds value for recycling. Interview partners recounted stories of contracts that failed because of the diverging expectations between the cooperatives and the contractors who were looking for a cheap way to dispose of waste.

Cooperatives often overestimate the benefits of collection contracts by underestimating their costs for labor, fuel, and machinery depreciation. The traces collected from trucks illustrated a set of constraints—spatial miscoordination, traffic and parking issues, uneven amounts of material at different pickup points—all of which diminish the seeming benefits of curbside collection. Several cooperatives holding service contracts did not reach minimum wage for their members.

None of the municipalities whose representatives we spoke with during our workshops had evaluated the cost of commercial services comparable to those offered by many of the cooperatives. According to estimates, similar commercial services would cost municipalities up to 300 percent more (Fergutz, Dias, and Mitlin 2011, 598), raising the possibility that cooperatives are significantly underpaid. Complaints about municipalities and companies that emphasize inclusive waste policies while underpaying waste pickers for their services were common in our interviews with cooperative members.

Comprehensive documentation of spatial operations can help a cooperative measure the actual value of the services it offers. These data can help them negotiate with municipalities and make the value of their services more visible to the public. These data also provide job training assistance for pickers who are becoming more engaged with computers through technology training programs and their own cell phone use. Cooperatives that have a shortage of labor can exploit data to emulate a foraging strategy, evaluating which clients to accept depending on the service costs and the value of the recovered material.

### How Much Formality Is Too Much?

The situation at O Verde é a Nossa Vida in Recife illustrates how formalization takes place in practice, highlighting a gray area between formality and informality. A small organization near Pró-Recife, O Verde é a Nossa Vida works as an association rather than a cooperative, which allows it to provide services to the public, but not work for profit.

Maturing from an informal group to an association gave the group access to partnerships. If it were a formal cooperative—a step that the city encouraged it to take—it could write invoices and sell directly to industries. However, the group would also have to recruit more members and would face higher costs and governance structures to maintain its status. The group decided not to take additional steps toward formalization.

In the current arrangement, the group maintains a formal relationship with the city and companies, but due to its status, it has to sell material to informal *Atravessadores*, or middlemen. The group has settled on a level of formality acceptable for now, but this approach limits the value of data that members collect and does not serve to further their independence from intermediaries.

### Forage Tracker's Failures and Lessons

In the Forage Tracker experiment, we used location-based media and participatory design to examine the spatial practices of organized waste picker groups, as well as their practices of information collection and exchange. In our work with cooperatives in São Paulo and Recife, we investigated different scenarios for making the following legible: their collection activities, coordination, and management, and more generally, the value of their service for the local neighborhood.

In the course of this work, we created the first maps of waste collection activities by Brazilian cooperatives, including the commuting distances of members. The results initiated an ongoing discussion about the role of data as well as their potential benefits and drawbacks. The collection map also proved to be of interest to people outside of the cooperatives.

Judged by the goal of designing a data collection system that would end up being used in a cooperative's day-to-day practice, the project was a failure. However, in terms of the goal to better understand the issues of informal waste management in conversation with the cooperatives, the project was more successful. The GPS traces were useful artifacts for exploring the facets of informal recycling. Traces from individual collectors revealed the highly idiosyncratic nature of manual collection based on shared (or unshared) knowledge.

The workshop in Recife produced a smartphone-based mapping prototype that cooperative members judged as valuable for managing collection fleets and clarifying the value of the services they offered. In daily operations, technical issues involving service coverage, hardware, and maintenance highlighted the shortcomings of the tool for everyday use. Nevertheless, the application was useful as a diagnostic tool for investigating information management and its challenges.

The situation in Brazil seems highly specific, but it also offers lessons for waste systems in the United States and Europe. Informal waste management exists anywhere in the world, including the rich capital cities of the West. Scavenging is not always the last resort for people surviving at the existential minimum. Many are what Scheinberg calls "part-time waste pickers" who collect recyclables for various reasons, to earn additional income or just because they enjoy it. In my Austrian hometown, the *Sperrmüll* dump that collects bulky waste and appliances is a fixed destination for tinkerers and explorers of all ages. At the MIT loading dock in Cambridge, Massachusetts, where decommissioned equipment awaits recycling pickup, one may find at any given time students rummaging for things they could use.

Informally organized recycling does not stop at the gates of cooperatives and associations. Many courteous residents around COOPAMARE participated by meticulously separating and cleaning their recyclables before bringing them to the cooperative. All these examples show that maintaining waste systems offers the potential for participation. How such participation is enacted, encouraged, and facilitated at the interface between citizens and governments will be the topic of part III of this book.

# Addendum: Structures of Brazilian Cooperatives

This addendum describes the organizational structures of the recycling cooperatives we worked with when conducting the Forage Tracker experiment in São Paulo and Recife (table A.1).

## Recycling Cooperatives in Greater São Paulo

The largest metropolitan region in South America, São Paulo is home to more than fifty recycling cooperatives, about twenty of which are organized in the local network Rede Catasampa and have partnerships with the municipalities of the region. The three cooperatives we worked with are sizeable and complex organizations, well connected with local governments and the waste picker movement.

### COOPAMARE (Cooperativa de Catadores Autônomos de Papel, Papelão, Aparas e Materiais Reaproveitáveis)

COOPAMARE is the oldest recycling cooperative in São Paulo; some of its members were cofounders of the national movement of waste pickers in 1999. COOPAMARE started as an informal association in 1986 before organizing as a cooperative in 1989 in a project initiated by the Catholic Church. It is located in the wealthy central district of Pinheiros, which offers advantages for collection but is inconvenient for most members who live in the poorer neighborhoods and endure daily commutes lasting several hours. Because the cooperative's central location is highly attractive to private recycling companies, COOPAMARE will likely face increasing competition.

The facility, which is owned by the prefecture and houses modern equipment such as sprinkler systems, is tucked beneath a freeway viaduct in a space initially occupied by the informal pickers. Rather than servicing a whole neighborhood, COOPAMARE operates a private business through

**Table A.1**
Overview of the investigated groups

| Name | Time visited | Form of organization | Founded by | Collection pattern | Role of government | Who buys material | Formalization model |
|---|---|---|---|---|---|---|---|
| Greater São Paulo area | | | | | | | |
| CRUMA Poa, SP | Jan 2011 | Cooperative | Collectors | Curbside, drop-off points | Provides facility, facilitator | Recycling companies | Service / hybrid model |
| Coop-Reciclavel Guarulhos, SP | Jan 2011 | Cooperative | City | Drop-off points, curbside | Provides facility, client | Recycling companies | Hybrid model |
| COOPAMARE São Paulo, SP | Jan 2011, Nov 2011 | Cooperative | Collectors | Clients, drop-off points | Provides facility, client | Recycling companies | Commodities model |
| Greater Recife area, Pernambuco | | | | | | | |
| COOCARES / COOREPLAST Abreu e Lima, PE | Jun 2013 | Cooperative | Collectors | Curbside | Client | Intermediaries | Commodities model |
| Pró-Recife Recife, PE | Jun–Sept 2013 | Cooperative | Collectors | Clients, institutions | Client and facilitator | Recycling companies | Commodities model |
| Verde é Nossa Vida Recife, PE | Jun 2013 | Association | Collectors | Curbside | Facilitator for private partnerships | Intermediaries | Commodities model |
| ARO Olinda, PE | Jun 2013 | Association | City | Curbside | City contracts collectors, oversight | City | Service model |

several partnerships with private and public entities. Up to three times a day, it collects from residential and business clients that include shops, apartment buildings, drop-off points, public institutions, and larger companies such as the supermarket chain Pão de Açúcar.

COOPAMARE is well equipped to implement NPSW. Companies have sponsored equipment, and NGOs are involved in educational and social projects. The cooperative receives municipal and federal grants, recently used to construct worker housing closer to the cooperative. During our fieldwork, the members generally had access to more material than they could process because of a shortage of labor, which was the main limiting factor to their business.

The cooperative uses a commodities model, selling material directly to industry. Many aspects of its operation are highly formalized. For example, a lawyer facilitates the monthly reports of processed material. However, there are many informal aspects as well. Truck routes are informally planned and documented in handwritten journals by the driver. The manual collectors working on their own do not coordinate activity with the driver. Instead, they collect, bale, and sell their own material to make more money. COOPAMARE owns two trucks, but during our site visits in November 2011, only one member had a suitable driver's license, which highlights the dilemma that members who acquire a license can quickly find employment at higher wages elsewhere.

### CRUMA (Cooperativa de Reciclagem Unidos pelo Meio Ambiente)

CRUMA, which is located in the city of Poá within the metropolitan region of São Paulo, is one of the oldest cooperatives in the city. It was founded in 1996 by the waste picker Roberto Laureano da Rocha and a few of his friends in an attempt to become independent from intermediaries. Like COOPAMARE, CRUMA was centrally involved in founding the national waste picker movement.

During our fieldwork, CRUMA consisted of forty-six members and collected eighty tons of recyclables per month from eighteen districts, which amounts to 10 percent of the total waste generated in Poá. The cooperative collects material from the curbside using a truck as a temporary collection point in a neighborhood and manual carts to visit individual households. CRUMA is also a community organization, operating a drop-off point for recyclables, running an e-waste center that accepts appliances, and serving as an educational institution for computer literacy using refurbished equipment. In response to the NPSW, the cooperative prepared a plan for extending its selective municipal collection.

The CRUMA facility is provided by the city, and the machines were acquired through various sponsorships. The truck was converted to run on vegetable fuel by the MIT research initiative Green Grease, one of several partnerships with universities (Colab MIT 2010). The cooperative works as a subcontractor for the waste management company that holds a city-wide collection contract. This is a source of discontent because CRUMA members have to make income from selling material rather than by collecting.

Because CRUMA receives grants from local and national governments for various environmental and social initiatives, the formalization model can be characterized as "commodities based." Despite grants and material support from the city, members are not compensated for collection and processing activities, and they do not earn minimum wage, making CRUMA operations not yet economically sustainable. Recently, CRUMA began to use the CataFácil software for managing its collected material and finances.

### COOP-RECICLÁVEL (Cooperativa de Materiais Recicláveis de Guarulhos)

With eighty members, COOP-RECICLÁVEL is a large cooperative that collects recyclables across the entire city. Started in 2003, it was inspired by CRUMA's model and founded by the municipality to implement a citywide curbside recycling system that processes paper, cardboard, plastics, glass, iron, aluminum, and e-waste.

The city plays a strong role in the daily operations of the cooperative, providing a well-equipped facility, two trucks, a driver, and fuel. The cooperative members who accompany the truck are responsible for sorting, separating, and baling material at the facility. COOP-RECICLÁVEL also operates voluntary collection points. Oversight, route planning, and data collection are in the hands of the municipality, which provides all necessary logistic services. The city's central role in daily activities indicates a service model.

The organizational form of a cooperative allows the selective collection of recyclables on narrow and partially paved streets, an environment where commercial hauling is practically impossible. The structure also allows the city to address social issues and take advantage of incentives provided by inclusive solid waste policies. Formally, the cooperative maintains leadership autonomy, with the city having no formal influence in management decisions. Nevertheless, a municipal official has an office on site.

## Cooperatives in Pernambuco

### COOREPLAST (Cooperativa de Reciclagem de Plástico LTDA) and COOCARES (Cooperativa de Catadores de Materiais Recicláveis Erick Soares)

The neighborhood Fosfato in the town of Abreu e Lima, a one-hour drive from Recife, is the home of two neighboring recycling cooperatives, each ranging between twelve and nineteen members and operating under joint leadership.

COOCARES, named after Érick Soares da Silva, an influential waste picker activist who died young, was founded in 2003 as an informal association during an organized protest of waste pickers on an open dumpsite at Inhamã. COOCARES focuses on cardboard, metal, and plastic that it sells to intermediaries. COOREPLAST, which was founded by waste pickers in 2004 and became a formal cooperative in 2009, specializes in plastic. Both cooperatives went through a business incubator program of the Federal Rural University of Pernambuco and received equipment from Petrobras.

The COOREPLAST facility is a significant obstacle to the group's development. Its area of four hundred square meters is split among several small buildings on different levels connected by narrow pathways. A machine for granulating plastic cannot currently be used due to lack of space, and PET is washed by the members in their own houses before processing. Separation takes place inside buildings, in small courtyards, and in the street. The COOCARES facility is slightly smaller but consists of one large space that is better suited to recycling. The workers, who had previously lived on the open dump, are less specialized and know how to collect and process all kinds of material, including textiles and shoes.

Together, the cooperatives operate two trucks and collect in six different neighborhoods of the town: Caetés I, Caetés II, Caetés III, Caetés Velho, Timbó, and Matinha. The drive to a collection site can sometimes take more than thirty minutes. The collection is organized in teams of six to eight collectors, using a truck and manual carts, covering about sixty streets per day. The truck serves as a temporary collection point in a central location. Teams of two or three collectors pick up material curbside and transport it to the truck using handcarts.

Surprisingly, neither cooperative collects in its own neighborhood, Fosfato, which is the territory of informal pickers who sell material to intermediaries. The national movement of waste pickers, MNCR, does not allow cooperatives to act as intermediaries that buy from informal waste pickers. The cooperative sees this regulation as counterproductive, since it could offer the informal pickers a better price for their material. Cooperative members explained that the informal pickers refuse to join the cooperative because they prefer regular hours and daily revenue to a monthly salary.

Many aspects of both cooperatives' operations are highly informal. Despite its status of a formal cooperative, COOREPLAST works with intermediaries, who offer lower prices but are in close proximity and accept material in smaller quantities. The cooperative has to sell material as quickly as possible because it has little storage space and lacks the financial cushion needed to wait for better prices.

COOCARES, on the other hand, sells about 60 percent directly to industry due to partnerships with Coca-Cola and the PET recycling company Frompet. The cooperative also provides services such as removing caps and labels from PET bottles—a process that currently cannot be accomplished by machines. Both cooperatives also trade material that they are not equipped to process with other associations. Members confirm that they could process and sell much more material if they had more space and more workers.

Both COOCARES and COOREPLAST use accounting and data collection, document the working hours of their members, and keep books on the materials collected and sold. During a 100-day program led in 2012 by the CATA AÇAO partnership,[1] both cooperatives learned bookkeeping and accounting, a practice still maintained more than one year after the project. Members take pride in their accounting skills. The books are not securely stored in the office, but placed prominently in the common room where everyone can see them. As the biggest benefit, keeping track of collected materials allows the cooperative to negotiate contracts with companies such as Frompet.

### Pró-Recife

Pró-Recife is Recife's largest cooperative, and it is a workplace for forty-one persons, mostly women. Located in the Boa Viagem district, the cooperative was founded in 2006 by a public-private partnership between the regional government, the AVINA Foundation, and the Walmart Corporation. The coop received machines, facilities, and training from this partnership.

Like COOPAMARE in São Paulo, Pró-Recife operates a private collection service with individual clients. They hold collection contracts for most public buildings and government institutions in Recife, and they provide collection services for large companies, supermarkets, and other generators of recyclable materials. Private collection creates logistic challenges, including traffic and driving restrictions, missed appointments, and a highly variable amount of material available at each site. Unpaved streets around the facility, which are regularly flooded and impassable during rains, are a serious service impediment.

E-waste is a major interest for the cooperative. Through its government contracts, it regularly receives waste equipment, but so far has been unable to make a profit from it due to the underdeveloped electronic recycling industry in Recife. A second issue stems from the reporting requirement of the NPSW, which demands that the cooperative document the source of the material before selling it at a profit. Despite its high intrinsic value, e-waste is currently less attractive than paper or PET.

Pró-Recife is one of the winners of the formalization process, representing a successful example of the commodities model. Facilitated by state and national policies, the cooperative has been able to secure many public and private contracts. By selling directly to the recycling industry, it bypasses intermediaries and receives higher prices.

Pró-Recife uses a computer for accounting and route planning. However, with the large and sparse collection area, monitoring the performance of collection and the yield per collection point is a major concern that remains to be solved. Since prices are negotiated with each client individually, better collection data could help the cooperative to increase its revenue.

### Associação de Catadores O Verde é a Nossa Vida

O Verde é a Nossa Vida, currently five people, has a small but well-equipped and well-organized space close to Pró-Recife in the Boa Viagem district. The group has existed for thirteen years and currently provides work for up to twenty employees. The association was founded in partnership between the city and a local packaging company that, unlike other groups, provides the facility and brings up to nine tons of material per month for sorting. The group received its current space in 2005 and formally registered as an association four years later.

Because the group is an association rather than a cooperative, it is allowed to sell services but not material, which it nonetheless sells to intermediaries informally. Since the association does not own a truck, it collects from the neighborhood around the facility using manual carts, usually three times a week. About six to seven tons of material are gathered from the street per month. Additional material collected from companies, stores, and condominiums makes the grand total a relatively modest fifteen tons per month.

Each of the collectors has a collection strategy. One interview subject collects only paper, cardboard, and PET from companies. Two other waste pickers collect PET mainly from residential buildings. Since many residents do not separate recyclables, they have to pick PET out of the waste.

In terms of information management, the association sends monthly mass-balance reports to the city detailing the amount of material collected. It sees reporting as an obligation toward the city that has little significance to its day-to-day work.

### ARO (Associação dos Recicladores do Olinda)

Olinda is a historic city that is designated as a UNESCO World Heritage Site. Its narrow, steep streets are not suitable for automated waste collection. Before the cooperatives were formed, three hundred waste pickers, many of them children, lived at an open dumpsite that the city has since closed.

To address the problems, the municipal sanitation department started a special project in 1998 called Projeto Meio Ambiente e Cidadania (PMAC).[2] The city hired pickers from the landfill in Água Fria to collect waste in the historic town, creating the Associação dos Recicladores do Olinda (ARO). It provided a space for separation and storage where pickers could work and sell material.

In 2003–2004, the city wanted to extend the program and provide equipment. However, the pickers were not convinced about the project's viability and kept returning to the open dump, which offered more material in close reach. The dumps were finally closed in 2007, and the NPSW of 2010 no longer allowed collectors to operate on dumpsites. At that point, the pilot program to recognize and train pickers to collect recyclables in the city gained traction (Prefeitura de Olinda 2010).

For waste management, Olinda is divided into ten areas, for which contracts to registered associations and cooperatives are provided. ARO collects recyclables and waste in the historic part of the city, with its steep and narrow roads. It educates residents about recycling and oversees the transportation, separation, and selling of material (Macedo and Furtado 2003).

Collection occurs three days a week, using a truck and vertical carts, which are better suited for the city's steep streets. ARO also collects during big events such as the carnival in Olinda, which creates a special logistic challenge. Recyclables are sorted at the facility. Every month, sorted material is purchased by the city on a per-weight basis. The city subsidizes the price for each ton of collected recyclables and waste, amounting to double what intermediaries pay.

ARO represents a form of a service model. The city is in charge of logistics, organization, and oversight. It provides the facility and covers costs for fuel, truck maintenance, and a driver. The long-term goal is to convert the group into an independent entity that can cover its own maintenance costs. Members have not yet reached minimum wage, so this is currently not realistic.

Data collection and oversight are conducted by the city. Officials monitor the work, weigh material, study waste composition, and administer an annual survey to measure citizen satisfaction, gathering ideas for improvement. A frequent survey response is to hire more pickers and extend collection. However, the city does not know the exact income of individual collectors. Because the city has not evaluated a commercial approach, it is not clear how costs for the association would compare. However, it can be assumed that a commercial service would be significantly more expensive for the city.

# III   Participation

# Prologue to Part III: Crowdsourcing Infrastructure

**Figure III.1**
Street sign in Cambridge, MA, requesting citizen participation in reporting maintenance problems with urban infrastructure.

Self-described as "investigative crowdsourcing," the German collective GuttenPlag Wiki produced extensive evidence of plagiarism in the dissertation of then-Minister of Defense Karl-Theodor zu Guttenberg (Anonymous and Kotynek 2013, 76). Interacting exclusively through pseudonyms, the group set up an online infrastructure to coordinate tasks and verify findings, establishing internal self-governance through techniques that exposed false findings by participants acting out of malice or overzealousness.

GuttenPlag is one of many examples of citizens organizing around a specific purpose and developing tools meant to hold a government accountable. Mechanisms for accountability are a central part of every democratic government that has to maintain legitimacy in the eyes of citizens. As political scientist Andreas Schedler puts it, "The great difficulty lies in this: you must first enable the government to control the governed; and in the next place oblige it to control itself" (1999). Accountability obliges one party to inform another about past or future actions and decisions, to justify these actions, and to accept sanctions in the case of violations (ibid., 17).

From the perspectives of both the system and the individual, accountability can be seen as the legibility of governance, made visible by transparency measures such as the U.S. Freedom of Information Act. As I show in chapter 5, official instruments of visibility often are not enough to establish the accountability necessary for a democratic society to function. When citizens perceive that government actions lack responsiveness or clarity, they may attempt to establish these values through official channels or by their own means, invoking rowdy shame campaigns or developing the tools and procedures that the government cannot or will not provide.

The GuttenPlag Wiki example illustrates the multifaceted roles that data-centric practices and visual representations play in community-driven initiatives. Citizen groups utilize visual representations for purposes that include recruiting volunteers, coordinating tasks, collecting data, analyzing evidence, and managing discussions. At the beginning of the GuttenPlag project, the text analysis and visualization tools were makeshift creations. I often describe such things as "dirty visualizations":[1] rich, cluttered, and less polished than professional design products, but essential artifacts of the operational process, containing many traces of the collaborative work that are no longer present in the final product.

As the GuttenPlag analysis neared completion, a more polished form of visualization became necessary for distilling the results for public discourse and building coalitions with journalists. Information designer Gregor Aisch created an elegant and complex visualization, documenting all identified instances of plagiarism in the dissertation.[2] This required the careful rewording of statements to make the findings quotable and to abstain from hyperbolic language that would compromise the group's role as a research platform.

The methods used in GuttenPlag are also applicable in other domains, including environmental monitoring. Traditionally, citizen-led initiatives depend on a well-organized community of volunteers who can make a considerable commitment in terms of time, knowledge, and resources. What can be effective for addressing local waste issues does not necessarily work at a larger scale when a large number of observations is necessary. Crowdsourcing initiatives try to overcome these limitations by reducing the necessary commitment to participation. The flood-mapping initiative Peta Jakarta accomplishes this by harvesting the Twitter network— popular among the residents of Jakarta—for reports of flooding and then invites the senders to contribute their observations to a shared real-time map.[3] International organizations including UN agencies are experimenting with crowdsourcing platforms to collect information for disaster and crisis response (Meier 2015) or to report illegal dumpsites (Fathih 2015).

Crowdsourcing can excel when measurement is a simple affair, as in the case of identifying and reporting software bugs by the users of the program. In the words of open source "evangelist" Eric Raymond, "Given enough eyeballs, all bugs are shallow" (Raymond 1999). Reaching the necessary number of "eyeballs" is not trivial, though, and it is the main reason why the majority of crowdsourcing projects fail. Crowdsourced monitoring initiatives also struggle with data quality issues, with data being notoriously biased, unreliable, and unevenly distributed.

In classic crowdsourcing, a participant solves an atomistic problem without necessarily seeing the larger purpose of the task, the contributions of other users, or how the collected data are used. Such a design may simplify participation for both contributors and users, but it also implies lower accountability on both sides. Environmental sensing projects frequently address this tradeoff by using a hybrid design that combines aspects of traditional community-oriented projects with the simplicity of crowdsourcing apps. Calibrating the visibility of users and their work both internally and externally is an ongoing process that will be further discussed in the third part of this book.

The design of civic interfaces plays a critical role in communicating any type of information. Given the complexity of infrastructure governance, interface design deserves particular attention for a reporting system in which citizens interact with governments and service providers. Using two case studies that look at mobile apps used within the Boston metropolitan area, chapter 6 examines how interface design affects the reporting of infra-structure issues such as potholes, broken streetlights, and graffiti. As I will show, interface design not only affects how problems are reported, but it also alters the perceptions of what constitutes an issue in the eyes of both citizens and the city.

# 5 Who Is Infrastructure? Participation in Urban Services

Pachube was an open Internet of Things (IoT) platform that allowed participants to "share real time environmental data from objects, devices and spaces around the world" (Pachube 2008). The shared sensor streams collected by amateurs and institutions included live data ranging from radiation and air quality measurements to the status of countless coffee machines in labs around the world.

Founded by the artist Usman Haque, the platform allowed individuals to connect networked sensors to the Internet and make their data accessible to others via its own application programming interface (API). Unlike earlier, constrained efforts such as personal automated weather stations connected by enthusiasts to websites such as the weather underground network,[1] Pachube was open and configurable for any kind of data stream. By 2011, the project had become popular enough to potentially become a Wikipedia for live sensor data, and the Pachube team was dreaming about a global sensor network maintained by hackers, artists, and urban activists. In an online conversation on the Pachube blog with the urban interaction designer Adam Greenfield, Pachube executive Ed Borden sketched a vision in which a volunteer collective shaped the future of urban data generation, in his words, "establishing their own standards and questioning the standards of others" (Borden and Greenfield 2011):

BigGov has become irrelevant in the public sector, eclipsed by someone with a supercomputer in their pocket, open source hardware and software at their fingertips, and a global community of like-minded geniuses at their beck and call: YOU. YOU are the Smart City.

… Adam, we need better verbiage here. What do we call this "citizen of the Smart City" and how do we make sure there are a whole lot more of them?

—Ed Borden

Greenfield responded with a different take on infrastructure (ibid.):

We call them "citizens," Ed.

... There are some things that can only be accomplished at scale—I think, particularly, of the kind of heavy infrastructural investments that underwrite robust, equal, society-wide access to connectivity. And for better or worse, governments are among the few actors capable of operating at the necessary scale to accomplish things like that; they're certainly the only ones that are, even in principle, fully democratically accountable.

—Adam Greenfield

Greenfield turned out to be right. Less than two months later, the Pachube platform was sold, and its new owners renamed it and turned it into a closed commercial service—a success for its original creators, but arguably a loss for the community of contributors. The fate of Pachube, its groundbreaking idea, growing community, infectious enthusiasm, and ultimately brief life as an open community illustrate a question that frequently arises when participatory online communities are compared with urban systems: can infrastructures be crowdsourced? The final part of this book examines the tensions between the acts of building and maintaining systems, between voluntary contributions and standardized maintenance protocols, and finally, between the role of mediating interfaces and technologies in this respect.

## User-Driven Infrastructure Paradigms

The notion that distributed, user-driven platforms such as Pachube can make traditional urban service provision obsolete is a central narrative of networked urbanism and appears in many forms. It claims that the decentralized, information-centric, perpetually adapting, and voluntary nature of online platforms offers a replicable model for the organization of urban services that is superior to traditional centrally managed systems. Critic Evgeny Morozov describes this position as Internet-centrism, which, in his view, is based on the mistaken assumption that the Internet is a coherent object with its own essential logic, rather than a heterogeneous assemblage of wires, protocols, and human practices that involve both centralized and decentralized aspects (Morozov 2014).

Two examples of user-driven infrastructures, embodying the "logic of the Internet," are frequently mentioned in optimistic accounts of possible technological futures: the world of open source software (OSS) development in general and the online encyclopedia Wikipedia in particular. They are often characterized as complex systems that miraculously work seemingly without central coordination. It is often overlooked that open source projects fail far more often than they succeed, and both examples involve complex governance structures and have many centralized aspects.

Nevertheless, open source projects, such as the Linux kernel developed under its "Benevolent Dictator for Life" Linus Torvalds, broke with the long-held assumption in software development that too many cooks spoil the broth. Despite relying on the work of amateurs, the software has the reputation of being stable and reliable. Software developer Eric Raymond attributes the success of the Linux kernel to what he describes as the "bazaar model": instead of releasing only clean and stable versions of the code base authored by a small team of experts, the project succeeded by frequently releasing imperfect versions, relying on the community of developers to find and fix problems (Raymond 1999). Raymond characterizes bazaar-style development as a perpetually unfinished process of small increments that involves a constant rewriting, reusing, and discarding of code. Users are treated as co-developers whose bug reports and suggestions are an integral part of the development process.

Considering that open-source software projects do involve not only creative problem solving but also many tedious and repetitive tasks, it may seem only a small step to applying similar principles to questions of infrastructure governance and maintenance. The communities of open software users need to be cultivated, but the constituents already live in the neighborhoods where they are affected by infrastructure deficiencies and could supply the local knowledge needed to resolve them. It might seem practical to address these issues through horizontal coordination and voluntary cooperation.

Such a position is abetted by technology advocate Tim O'Reilly, who offers OSS development principles as an alternative to bureaucratic decision making: "Open source software projects like Linux and open systems like the Internet work not because there's a central board of approval making sure that all the pieces fit together but because the original designers of the system laid down clear rules for cooperation and interoperability" (O'Reilly 2011).

The alleged advantage of clear and transparent rules of code over messy politics is a recurrent theme in tech literature. This characterization, however, ignores the variety of governance models hidden under the broad umbrella of *open source*: its vastly different approaches to motivating contributors, resolving conflicts, and planning future directions. Open source projects are not always exemplars of democratic and decentralized governance, but can fall anywhere on a spectrum from a dictatorship to anarchy. Projects such as the Linux distribution[2] Debian have a constitution and elected leaders, while others have no governance structures at all. Many large projects such as the Linux kernel, Wikipedia, or the content management system Drupal have "Benevolent Dictators for Life" (BDFL). Within this group, some projects such as Wikipedia have detailed policies for decentralized conflict resolution, while others, such as the Linux kernel, are strongly centralized with the BDFL, usually the initiator of the software project, making strategic decisions that settle all disputes (Fay 2012).

The application of open source models to urban governance can mean, quite literally, using and developing OSS for cities. The nonprofit organization Code for America's mission statement declares, "We build open source technology and organize a network of people dedicated to making government services simple, effective, and easy to use" (Code for America 2016). Former president Barack Obama's Memorandum on Open Government calls for applying new technologies for managing and distributing information for purposes of government transparency (Obama 2009). Five years earlier, the city of Munich started the migration of all its software systems to open source software (Casson and Ryan 2006).

More often than not, however, "open source" is used metaphorically for describing distinct values and principles. For sociologist Saskia Sassen, open source embraces incompleteness and thus resonates with her understanding of cities, which are constantly remade, not only by powerful actors, but also by citizens, who may resist through their practices (Sassen 2011). In a similar vein, anthropologist Alberto Corsín Jiménez frames urban infrastructure not as a finished system, but as "a prototype, whose main quality is its permanent 'beta' condition" (Jimenez et al. 2014). By laying out the affordances and deficiencies of infrastructure systems open to the public, open source has a justice dimension becoming a means toward what he calls the "right to infrastructure" (ibid.).

Incompleteness, however, can only be seen as a positive value in combination with a capacity for continuous improvements. In the words of Beth Noveck, lawyer and head of the White House Open Government Initiative: "Whenever we confront a problem, we have to ask ourselves: How do I parse and distribute the problem? How might we build feedback loops that incorporate more people" (Lathrop and Ruma 2010, 49)? To facilitate distribution and asynchronous collaboration, several tools used for managing open source projects have been reimagined for urban governance. This includes version control systems (VCS), platforms that document every change to the codebase and facilitate communication and deliberation grounded in the activity of coding (Fuller and Haque 2008). Through VCS, bug trackers, and other web platforms, deliberation can happen asynchronously, and every change to the codebase can be reversed.

This space of asynchronous collaboration, where every action within the platform is legible in its historical context, enables a specific culture of participation. As anthropologist Chris Kelty describes, a substantial part of the work of developer communities is dedicated to building and refining not only the software to be developed, but also the very tools and modes of communication necessary for coordinating the community—in his words, a "recursive public" that constantly rebuilds itself (Kelty 2005). Within this recursive public, individuals do not assume exchangeable roles as equal but generic citizens, but become active in their own areas of expertise—a phenomenon that human–computer interaction researchers Stacey Kuznetsov and Eric Paulos describe as the "rise of the expert amateur" (Kuznetsov and Paulos 2010).

The qualification "expert" is relevant in this context. The widespread assumption that the impressive products of online collaboration such as the Linux kernel are the result of hobbyists with too much time on their hands is contradicted by empirical studies that reveal that many participants are self-selected professionals and experts (Brabham 2012). According to a report by the Linux Foundation, only about 15–20 percent of the Linux kernel is written by independent developers; the rest is contributed by companies such as Intel or IBM (Kroah-Hartman, Corbet, and McPherson 2008).

The lesson drawn by initiatives such as Code for America is that not everyone needs to get involved—after all, only a tiny fraction of Wikipedia users also edit or add content. Instead, the goal of these initiatives is to reach those potential expert amateurs who may contribute to platforms such as Pachube, can build systems for measuring radiation, or have an intimate knowledge of vacant lots in their neighborhood.

### Decentralization versus the Ideal of Integrated Infrastructure
The rhetoric celebrating the advantages of decentralization and co-production does, however, raise a question: is centrally managed, efficient infrastructure really such a bad thing? More precisely, when did it start to be seen as such?

Decentralized models for building infrastructures certainly predate the Linux kernel. Historian of technology Thomas Hughes illustrates the transition from modernist, top-down infrastructure development with its central hierarchies to the postmodern modes of planning in heterogeneous networks through the example of Boston's Central Artery/Tunnel (CA/T) Project. The highway tunnel project in central Boston, completed in 2007, replaced an elevated inner-city highway built during the 1950s and took more than twenty-five years to complete. Unlike fifty years earlier, the main challenges in the CA/T Project were no longer technical and logistical, but social and organizational. Without a single central planning authority, the interests of many different actors had to be negotiated—federal laws were enacted to secure funding, firms founded in joint ventures, environmental assessments conducted, and especially important in the last phase of the project, pressure from media and interest groups negotiated (Hughes 1998, 197ff). In contemporary infrastructure planning and governance, urban systems are rarely operated by a single entity, but typically by a hybrid network of actors with different relationships and dependencies: various governmental agencies, utility companies, financial institutions, and other interest groups.

By the late 1960s, most cities in developed countries were connected to the four essential public services—water, transportation, sanitation, and electricity—which were built, owned, and operated by the public. Geographers Steve Graham and Simon Marvin term this form of uniform, standardized, and universal service provision the "integrated infrastructural ideal" (Steve Graham and Marvin 2001, 73). Centralized infrastructure development has many advantages. The high initial costs create natural monopolies and make it difficult for private actors to compete. Economies of scale increase efficiency with the size of the system, and the consensus that these services constitute public goods makes governments their natural custodians.

As chronicled by Graham and Marvin under the label of "splintering urbanism," the urban infrastructure landscape became fragmented during the second half of the twentieth century as the public hand largely withdrew from service provision and privatized central utilities. They describe a number of causes responsible for its demise. The economic crises of the 1970s and the departure from the Keynesian welfare state left many public projects underfunded, diminishing the quality of service. The changing political economies of globalization further weakened the role of the state as the main provider of infrastructure. Finally, the departure from the modernist command-and-control planning paradigm made infrastructure projects increasingly complex and expensive, a development accelerated by social critiques and fierce civic opposition to federal infrastructure projects conducted under the urban renewal program in the United States, which involved large-scale eminent domain, demolition, and resettlement (Steve Graham and Marvin 2001, 91ff).

Graham and Marvin diagnosed increased inequalities along spatial, economic, and social dimensions due to the fragmentation of infrastructural networks. In the process of privatization, the large services became unbundled and marketed in different locations as separate services with different prices for different user groups. This fragmentation, reinforced by the economic logic of private service provision, creates winners and losers. Economically disadvantaged groups or areas often end up without service, or they have to pay a higher price for it (ibid., 284).

### The Legibility of Infrastructure Services

The described processes of infrastructure decentralization affect how users perceive and engage with a system. The deteriorating condition of urban infrastructure, power blackouts, and inequalities of service provision draw attention to infrastructure and make these systems more visible. The same conditions also make infrastructure more participatory—though, alas, unaccompanied by the fanfare of engagement and empowerment. Brittle and unreliable systems, not just in the Global South, require the increased involvement of their users, improvised solutions, and informal coping strategies (Graham 2009, 144). Paradoxically, the abundance of competing services can have a similar effect. Consumers become more actively involved with an infrastructure by having to choose among services whose relevant differences are not immediately obvious (Steve Graham and Marvin 2001, 148). Many services require a higher level of infrastructure literacy on the side of the user than they used to require in the past. In the example of waste management, the residents of Seattle require a basic understanding of recycling processes to decide whether the greasy pizza cardboard box or the plastic bag should go into the recycling bin or the trash can.

The line between service provider and consumer is becoming blurred as users increasingly get involved in service provision. What started with subcultures such as the home-power movement has become commonplace (Tatum 1992). In many countries, households operating photovoltaic panels receive incentives or compensation to feed extra electricity back into the grid. Even on a more mundane level, more and more tasks that were previously in the domain of the service provider are shifted to users. This can happen visibly, as in the case of users having to manage bank accounts, or invisibly, as in regard to smart meters, which shift some of the grid's operational logic to users and collect load data without them being necessarily aware.

The same processes of decentralization also change the perspective of the provider. For private utilities, the profitable operation of an urban service requires making service consumption legible. In a market of competing infrastructure services, it is difficult to cover investment costs through regular service rates, since these tend to move toward the marginal cost of delivering the service (Frischmann 2012). A solution to this dilemma is to measure service consumption at a granular level and divide the rates and conditions of service delivery into different segments. For example, to offer competitive rates for cell phone service that would otherwise not cover the costs of installing and maintaining base stations, a provider can charge more for text messages or prepaid phone services, which are used by demographics that are of less interest to the provider. Information technologies and sensor networks give providers the fine-grained measurements necessary to enable the billing of service consumption and the feeding of user consumption data into dynamic pricing models. By introducing multi-tiered service costs that depend on local conditions, providers can take advantage of unbundling and service fragmentation to recuperate their investment costs. As Graham and Marvin argue, multi-tiered service provision and unbundling introduce inequalities to service delivery across user groups, services, and geographies: by charging users of prepaid services a higher rate than subscribers, introducing different rates for equivalent services, or offering a service only in profitable areas (Steve Graham and Marvin 2001).

### Smart Cities—Infrastructure in the Background

An electricity network in which every household is connected through a smart meter that measures and submits real-time usage data offers more possibilities to the provider than just cleverly segmenting services for profit. By dynamically adjusting service rates to individual consumption relative to the overall system load, the provider can directly influence the user's behavior in order to balance the load on the system. As a result, less power is needed during peak times, and the system becomes more efficient—the grid becomes a dynamic feedback system that can be optimized. Furthermore, based on usage patterns in collected data as well as interactions with other networks and external events, future states of the system can be anticipated. This is, concisely, the promise of the "smart city," a term that appeared as early as 1992, and was further discussed by Bill Mitchell and others in the following decade (Gibson, Kozmetsky, and Smilor 1992; Mitchell 1995, 41; Stephen Graham and Marvin 1996). The idea originated in earlier work on urban cybernetics, which conceptualizes the city as a dynamic system in which all actors, including its planners, constantly adapt to each other's actions, never reaching a static equilibrium (Forrester 1970; Goodspeed 2015). To accomplish this level of granular legibility, a smart city involves geographically distributed sensors and information infrastructures to measure the state of water, sanitation, electricity, transportation, healthcare, or policing.

Smart cities occupy an ambiguous place in the context of the decentralization, participation, and legibility of infrastructure. Smart cities are both agents and products of infrastructure privatization and fragmentation by involving IT companies such as IBM, Cisco, and Siemens in the management of public infrastructure. At the same time, its advocates strive for convergence instead of fragmentation, promoting the integration of diverse services in a unified model operating behind the scenes. The synoptic vision of urban processes, however, remains reserved for the administrator. The resident of a smart city is blissfully unaware of disasters mitigated, traffic jams averted, energy saved, and crimes prevented. Outspoken critic Adam Greenfield characterizes the smart city as a clinical, reductive, and generic model situated in generic space and time and predicated on notions of optimization and objectivity that are inappropriate for dealing with urban complexity. Greenfield argues that the idea of the smart city is based on a paradigm of seamlessness that is ultimately unachievable: "When systems designed to hide their inherent complexity from the end user fail, they fail all at once and completely, in a way that makes recovery from the failure difficult" (Greenfield 2013).

Indeed, IBM's white paper outlining the company's "Vision of Smarter Cities" did not consider citizen participation; instead, it offered a technocratic perspective that refers to citizens only as the recipients of services that enable them to enjoy a high quality of life (Dirks and Keeling 2009). Urbanist Anthony Townsend observes: "The technology giants building smart cities are mostly paying attention to technology, not people, mostly focused on cost effectiveness and efficiency, mostly ignoring the creative process of harnessing technology at the grass roots" (Townsend 2013, 118).

## Civic Technologies—Infrastructure in the Foreground

While the smart city model frames citizens primarily as consumers and passive sources of information, the contrary model of civic technologies emphasizes the agency of the individual, co-production, and creative appropriation. As critics like Townsend and Greenfield argue, instead of instrumenting the city with soon-to-be-obsolete sensors, technology can be used in more inclusive and participatory ways, leveraging local knowledge and ubiquitous technologies like smartphones. While smart city projects aspire to provide comprehensive solutions to general problems, civic technologies present themselves as more nimble and incremental, focusing on the nuts and bolts of local issues. However, just like the smart city, civic technologies are based on a data-centric ideology, grounded in the belief that improving coordination and access to relevant information is the key to solving urban problems. Nevertheless, urban problems such as segregation and gentrification cannot be solved solely through information—on the contrary, they seem to be exacerbated by information-saturated real-estate markets. I will address a number of broader critiques of civic technologies in the epilogue to part III.

Over the following sections, I will examine and compare different instances of a prototypical civic tech application that embodies the proclaimed values of openness, participation, and engagement—citizen feedback applications mediate citizen-government interactions and facilitate information exchange, feeding local knowledge back into the governance process. The following case study will show a more nuanced picture of civic technologies in action. The central concern is not the fact that something is made legible; rather, the issue is the effect of different approaches to establish legibility for public discourse on infrastructure governance.

The recent history of 311 systems in the United States illustrates the evolution of a feedback mechanism from a simple method of logging complaints and nonemergency incidents to an ambitious tool for civic engagement using telephone helplines, websites, and smartphone applications. In the context of this investigation, 311 systems are an interesting exemplar because they establish infrastructure legibility in two directions. First, they afford governments a detailed reading of the situation on the ground and the attitudes of constituents. Second, they make the activity of the local administration legible to constituents. By focusing on actual incidents, they offer a window into the material reality of infrastructure maintenance. Legibility in both directions is mediated by communication technology, which can act as a filter, an amplifier, a resonator, and a switch. Precisely how the interactions between citizens and governments are shaped through these interfaces will be the subject of the following case study.

## A Brief History of 311 Systems in the United States

The history of 311 citizen feedback systems in the United States is a story of growing ambition, provisional prototypes, and incremental improvements.

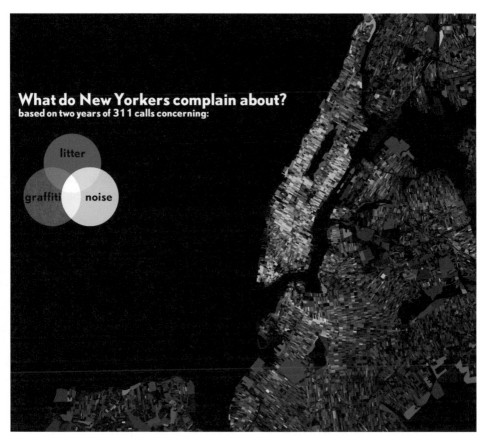

**Figure 5.1**
Map showing the spatial distribution of citizen complaints via New York's 311 helpline with respect to three different complaint types, mapped to the RGB color channels: noise (green), graffiti (red), and litter (blue). The resulting color allows an estimation of the proportional distribution of each complaint type in different parts of the city. Visualization by the author.

Within a decade, what started as an attempt to relieve the load on emergency call centers and provide better access to services has become a primary means for data collection about the condition of urban systems, a tool for public accountability and citizen engagement, and a conduit for government and citizen cooperation with infrastructure maintenance.

By the late 1980s, the police, fire, and medical emergency number 911 had become so popular for nonemergency requests that the call volume became a headache. Public management scholar Malcolm Sparrow and his coauthors quote a police executive who declared in 1985, "We have created a monster" (Sparrow, Moore, and Kennedy 1992, 105). The exploding number of cell phones only aggravated the issue.

To address this situation, in 1997 the U.S. Federal Communications Commission (FCC) designated the short code 311 for requesting nonemergency public services (Flynn 2001; FCC 1997). Some cities, including Buffalo and Baltimore, kept the nonemergency calls within the purview of policing. Other cities, among them Dallas and Chicago, integrated 311 call centers into local government (Mazerolle et al. 2002). Chicago launched its 311 community response system in January 1999 because of the urgent need to replace a non-Y2K-compliant mainframe system (City of Chicago 2013).

In 2002, then-Mayor Michael Bloomberg announced a 311 system for New York City as his first major policy initiative. At that time, twelve call centers served more than forty city agencies, often with significant overlap in competencies. Set up as part of the Office of Operations, the NYC311 call center was initially staffed by 300 operators who entered requests into a service management system used for scheduling department tasks. During the start-up phase, analysts and engineers continuously revised the service category assignments, protocols, and database structures used to parse and route the incoming requests. In 2009, NYC311 offered a web interface for submitting and tracking reports. By 2011, it was handling twenty-two million calls annually, more than the combined total of the next largest twenty-six cities with 311 call centers (New York City 2013).

After the launch of NYC311, the emphasis shifted away from the initial goal of load reduction. Although early experiences in Baltimore—where the police remained in charge of the nonemergency number—had shown a decrease in the volume of emergency calls (Mazerolle et al. 2002), a reduction did not occur when 311 calls were handled by the city (California, Department of General Services 2000). In New York, the goal was recast as simplifying access to city services for a multilingual constituency while simultaneously evaluating performance and increasing accountability (Cardwell 2002).

The focus on accountability included not only the city's responsibility toward its constituencies, but also the horizontal and vertical relationships between government entities. An NYC311 technician noted in an interview with the author that former Mayor Bloomberg was a frequent 311 caller; he wanted to observe how requests were handled by different departments. Unlike earlier systems, NYC311 assigned a unique ticket number that allowed each issue to be tracked from request to resolution. This approach, referred to as "constituent relationship management," was modeled on cus-tomer relationship management (CRM) systems used by large companies to track customer requests and schedule response tasks.

The data held in call-center CRM systems were useful in different respects. Complaints offer feedback on urban problems through the eyes of citizens. With their urban issues digitized, georeferenced, and categorized, city managers started to view 311 call data as valuable resources for measur-ing the quality of services. Because citizen calls represent self-reported data instead of random samplings, CRM data are biased in many different ways and present a challenge to scientific analysis. Although the large data vol-ume allows controlling for suspected biases through statistical modeling, the relationship between self-selected participants and the general popula-tion is poorly understood and limits the generalizability of results. This is, of course, a problem that is not restricted to citizen reports, but affects all large-scale data sets that are assembled from user-contributed sources (Mayer-Schönberger and Cukier 2013, 39; Hargittai 2015).

Nevertheless, citizen calls captured issues that would otherwise have gone unreported. For instance, the city of Chicago used citizen reports for combating bed bug infestations (Gabler 2010), and New York's 311 data were instrumental in identifying an episode of air pollution by locating the source of a mysterious smell reported by residents in a particular neighbor-hood (Johnson 2010). Reported issues were also used to create econometric models that tracked the perceptions of neighborhood characteristics over time, including empirical tests of the infamous "broken windows" theory (O'Brien, Sampson, and Winship 2015).

### Early Online Systems: Mashups and Civic Amateurs

With 311 call centers recording incidents based on operators asking ques-tions, online submissions through visual interfaces represented the next stage of citizen feedback systems. In the first "web map mashup," amateurs reverse-engineered some code of the Google Maps service. In January 2005, housingmaps.com georeferenced rental listings, making them searchable by location (Singel 2005). In May of that year, ChicagoCrime.org offered georeferenced crime reports scraped from police logs (Holovaty 2005). Do-it-yourself cartography officially emerged that summer when Google released an API for its map service, allowing anyone to create online maps using his or her own data (O'Connell 2005).

In July 2005, public advocate Andrew Rasiej launched WeFixNYC.com,[3] which let users upload photos of potholes to a photo-sharing website and georeference them in Google Maps (Shulman 2005). Unlike the earlier mashups, WeFixNYC invited users to create their own data. Driven by the emergence of smartphones with cameras and location sensing, similar services followed. In February 2007, the first mature system for submitting urban incidents started operation in the UK, followed in 2008 by the U.S. platform SeeClickFix.

By 2009, users could easily enter a location and a description in a smartphone app, turning the job of reporting into little more than taking a picture and assigning it to an incident category. The mobile apps for SeeClickFix and FixMyStreet were released. In a novelty for local government, the city of Boston's office for New Urban Mechanics, in collaboration with the mobile startup ConnectedBits, released the reporting app CitizensConnect. The apps sponsored by municipalities sought to improve the city's understanding of how citizens use services. Analysis of 911 calls had revealed that a remarkably large proportion of calls came from only a few addresses (Sparrow, Moore, and Kennedy 1992, 105). Chris Osgood and Nigel Jacob from New Urban Mechanics have noted in a discussion at Northeastern University, "We do not need to receive a report from everyone; we want to find the people who submit a lot of reports."[4]

By this point, many local governments had started to embrace civic technologies, developing online tools and mobile apps or licensing platforms such as SeeClickFix. The narrow focus on data generation began to shift to a more ambitious goal of involving citizens in infrastructure services, engaging them as stewards of their environment. In one of the first papers to analyze a data set of citizen-submitted incident reports, researchers Stephen F. King and Paul Brown lay out a roadmap as follows:

In the first stage, local government deploys ICT to improve information provision to citizens and to enable transactions with citizens to be conducted electronically ("the Responsive council"). In the second stage the data generated by these interactions is analysed by local government to generate insight into service use and future demand ("the Insightful council"). In the final stage, citizens take the lead and, through sharing information with each other and with local government, become active participants in service design and delivery ("the Insightful citizen" stage). (King and Brown 2007)

### Open311 and Open Data

Mobile citizen feedback apps fall within the field of volunteered geographic information (VGI) systems, which include participatory mapping projects like OpenStreetMap as well as disaster relief and accountability platforms like Ushahidi (Goodchild 2007). Despite the large number of reporting apps and platforms, most share similar functionality: take a picture, verify the location, select an incident category, and enter a text description. But with many cities and developers building similar tools, issues of interoperability and standardization have emerged.

Open standards are an important factor for the widespread adoption of e-government tools. The nature of open standards makes it possible to use a broad range of clients, platforms, and interfaces while generating machine-readable data that can be incorporated into other applications. The first Apps for Democracy Contest held in Washington, DC, in 2009 introduced an open standard for incident reporting, Open311, which aimed to put cities in a position to share data and quickly implement a feedback system based on interfaces that could be improved by outside developers.

Another requirement for civic technologies in the digital age is the open data principle, the public provision of government data in a structured and machine-readable format for unrestricted use (Lathrop and Ruma 2010). The U.S. Freedom of Information Act (FOIA), specifically in the instance of the Government in the Sunshine Act of 1976, granted citizens access to information unless restricted for privacy or national security reasons. However, FOIA requests can take several months to process. The photocopied or scanned pages released for the request are not machine readable and therefore of limited use for computational analysis, requiring meticulous labor to render the data in a digital format.

Open data improves data exchange, allowing developers to build applications that utilize information freely. Commercial dining guides, for instance, can benefit from restaurant inspection reports. Traffic guides may incorporate public traffic and weather information. The opening of accurate GPS data for civilian use in May 2000 has nurtured an industry that offers a range of location-aware services for portable devices. For governments, implementing open data requires a slow coordination of heterogeneous entities. As New York's Chief Data Analytics Officer commented during the Open Data Conference 2015 in Ottawa, "Convincing agencies to share their data is like pulling teeth."

## The Evolution of Accountability and New Public Management

Beyond a means for sending service requests, citizen feedback platforms are accountability mechanisms that allow users to follow up on issues directly with city workers. This two-way connection turns infrastructure governance into an interactive process, a conversation. In their passionate case for civic technologies, Goldsmith and Crawford reference citizens who say that calling the 311 hotline makes them feel like they are complaining, whereas reporting apps make them feel that they are helping (Goldsmith and Crawford 2014). The authors argue that such tools also improve local governments, since they introduce a new form of accountability that focuses on results rather than processes. Since all interactions with citizens and their outcomes are reflected in public data, civil servants are judged by the public based on these results rather than on compliance with internal guidelines. In their diagnosis, process-centric accountability is responsible for what they describe as a current crisis in local government: a general ossification.

This argument, however, has a longer history and echoes central notions of the New Public Management (NPM) doctrine, which similarly called for a redefinition of "accountability," from processes to results. Rooted in the conservative reorganizations of the public sector in 1980s Britain, NPM promotes a business-oriented model of governance that involves replacing bureaucratic accountability mechanisms with what public administration consultant Richard Boyle calls "post-bureaucratic control mechanisms," which involve contracts and partnerships with private firms, continuous performance monitoring, and private sector management techniques (Boyle 1995).

Shortly after the collapse of the Soviet Union, management consultants David Osborne and Ted Gaebler called for an "American Perestroika" of public management (Osborne and Gaebler 1992). Characterizing public service provision as slow, inefficient, and fundamentally outdated, the authors postulate the following principles to improve public management:

• Government should be a catalyst, not a service provider; it should steer, not row.
• Government should be owned by the community; the community should control, rather than receive services.
• Agencies should follow a defined mission, rather than rules and regulations.
• In evaluating services, one should measure the outcomes, rather than the inputs.
• Services should be driven by the needs of customers, not the bureaucracy.
• Governments should decentralize authority and embrace participatory models.
• Governments should use market forces to achieve change.

Despite the ostensible emphasis on empowerment and participation, the benefits and success of NPM remain highly controversial. Public management theorist Kulachet Mongkol summarizes the various critiques of

NPM in three broad points (2011). He calls the first the "paradox of centralization through decentralization": the introduction of managerial and market-oriented principles under the banner of decentralization has led to a centralization of decision making by concentrating authority in the hands of a few public managers. This concentration is problematic since, as he argues in his second point, private sector managerial approaches are not directly applicable to the public sector: they tend to emphasize simple solutions for simple problems and fail to account for the most basic requirements of democratic governance (Drechsler 2009). Finally, the emphasis on measuring performance over process necessitates a new bureaucratic apparatus for conducting assessments that can introduce its own problematic ethical standards and incentives (Mongkol 2011).

While traditional public accountability instruments are designed to prevent waste and corruption, NPM shifts the emphasis toward measuring service quality. Paraphrasing political scientist Christopher Hood, NPM constitutes a shift from public accountability to accounting (Hood 1995). The claimed benefits of key performance indicators (KPIs) on the quality of service provision remain controversial. Particularly in the domain of law enforcement, performance metrics that reward officers based on their number of arrests have come under criticism (Mazerolle et al. 2002). Even in the less contentious area of infrastructure maintenance, KPIs can limit the discretion of public officials, while it is not always clear whether metrics such as the number of filled potholes are an accurate proxy for service quality. Introduced to facilitate decentralization, KPIs paradoxically introduce centralization by requiring comprehensive information infrastructures for measuring service quality.

Many of Osborne and Gaebler's principles resonate in contemporary visions of a participatory, user-driven infrastructure that emphasizes the roles of open source, public engagement, and empowerment: the diagnosed failures of the public sector and the capacity of digital technology to resolve these failures by empowering constituents to take matters into their own hands. Yet citizen feedback systems and open data initiatives are not in all respects aligned with the goals of NPM. By creating new public services and platforms, current digital initiatives depart from the NPM imperatives of cost efficiency and the devolution of public utilities. Their information infrastructures are not necessarily intended to measure performance, but also to collect and integrate local knowledge. Citizen feedback systems are not instruments to limit the role of government, since managing requests is a considerable additional burden. Nor do they typically reduce costs for a municipality, breaking with the central NPM paradigm of economic efficiency. As Nigel Jacob commented in a discussion:

Citizens Connect is extra work for everyone. It does not save money, and nobody has checked this, because money is not the metric. Like in policing, the appropriate metric is not the number of arrests, but the subjective feeling of safety; quantitative metrics can lead to perverse incentives. Citizens Connect is really about engagement. Inclusive language and the perception of value are important. Our conceptual model is different from that of the *Smart City*, where efficiency is central.[5]

## Mechanisms of Accountability

Traditionally, accountability is a vertical relationship between citizens and elected officials or between a principal and a subordinate. Horizontal relationships of accountability also exist, for example, between agencies, in scientific peer review, or professional evaluations and appraisals.

Accountability mechanisms for urban service provision can be implemented either by the "short route," which directly connects citizens to service providers to resolve issues, or the "long route," which uses a public authority as an intermediary. Studies suggest that the long route, allowing municipalities to enforce the contractual compliance of the utility, is more effective in getting citizens' complaints resolved (Fox 2015).

Citizen feedback systems often involve more complex relationships of accountability, both horizontal and vertical, formal and informal, and always involving a large number of stakeholders. Many citizen feedback systems can be described as social accountability initiatives that aim to establish community-driven approaches that keep power holders accountable (Joshi and Houtzager 2012). While these initiatives often start with the community itself, they can also be spearheaded by an institution. International lenders like the World Bank promote social accountability as a way to combat corruption and to monitor how their funds are used in infrastructure projects.

Mechanisms of social accountability can be both formal and informal, operating through the judiciary or public pressure. When the formal mechanisms such as elections or court systems fail or are unavailable, accountability initiatives resort to informal channels such as media campaigns and protests. In the absence of enforcement and formal sanctions, constituents might turn to tactics of naming and shaming, what sociologist Naomi Hossain describes as "rude accountability" (Hossain 2010).

Social accountability is increasingly employed in development projects for improving urban services. Service providers can be held to account more effectively by international lending institutions, if the beneficiaries of services are directly involved in monitoring (Cavill and Sohail 2004, 155). In this view, social accountability could help prevent the misspending of public funds as well as make services more equitable to those who otherwise have no voice. By bringing infrastructure governance into the foreground, the approach can improve services and increase the perception of urban services as a public good.

Digital information technologies can play many different roles in helping social accountability projects make sense of infrastructure and public services (Offenhuber and Schechtner 2013). Sociologists Eric Gordon and Paul Mihailidis describe "civic media" as "the mediated practices of designing, building, implementing, or using digital tools to intervene in or participate in civic life" (Gordon and Mihailidis 2016). When supported by local governments, these practices reintegrate functions into the governmental sphere that were neglected under NPM. This model is sometimes described as Neo-Weberian, since it reaffirms the central role of the public sector in solving urban problems (Pollitt and Bouckaert 2004; Dunleavy et al. 2006).

Digital platforms have been used to document corruption or monitor elections. An initiative to map violent incidents after Kenya's disputed presidential election of 2007 led to the development of the popular crowdsourced mapping platform Ushahidi (Okolloh 2009). In cases such as these, civic technologies depend on the support of a dedicated community for development and to protect them from destructive forces and cooption (Zittrain 2008).

Because social accountability initiatives require a system of governance that respects the role of the community, they must interact with formal mechanisms of enforcement, bringing them to the limits, if their scope does not include questions of procurement and contract negotiations. Service providers are not accountable in the way that public officials are to their constituents—eventually, they are only responsible for complying with the conditions specified in their contracts.

## Civic Technologies in Action

Smart city visions and civic media practices are often described in the dichotomy of top-down versus bottom-up: the urban manager versus the citizen-activist, central control versus decentralized organization, or the private versus the public good.

A pure version of a smart city might resemble the authoritarian dystopia caricatured in Jean-Luc Godard's 1965 film *Alphaville*.[6] At the same time, attempts to create a "responsive city" risk turning a city into something that merely reacts to requests from those who are the most vocal. Address-ing citizen requests can bind scarce resources, and short-term fixes can replace long-term strategic planning. In both cases, increasing amounts of public data heighten the temptation to read these data sets as the reality of urban infrastructure problems.

Distinctions of *top-down* and *bottom-up* are evocative metaphors but not always useful categories for understanding infrastructure because they tend to obscure the multifaceted nature of large socio-technical systems. A closer look at the technologies reveals that these dichotomies are not as clear cut as they might seem. Similarly, even smart city solutions are not as mono-lithic as frequently presented by both their supporters and critics.

The engagement of the citizen has also changed. As sociologist Michael Schudson explains, the ideal of democratic decisions made by fully informed citizens is no longer attainable—if it ever was. Instead, Schud-son sees the rise of the "monitorial citizen" who concerns himself or her-self with selected issues and possesses an unfocused awareness of what's relevant to his or her interests, who "scans (rather than reads) the infor-mational environment," and is ready to mobilize when alerted (Schudson 1998). As the next chapter shows, civic tech tools in the form of feedback platforms that share functionality have emerged from both top-down ini-tiatives and activist projects. They handle the same issues and share many similarities, but do so in ways that represent governments and citizens dif-ferently, their interface designs affecting the perception of what is and what is not a problem.

# 6 The Urban Problem at the Interface: Reading Governance

Civic feedback systems have received considerable attention as instruments of accountability, data proxies for studying urban phenomena, tools for civic engagement, forces of anti-corruption, and conduits for participatory infrastructure governance (Desouza and Bhagwatwar 2012; Gordon and Baldwin-Philippi 2013; O'Brien, Sampson, and Winship 2015; Zinnbauer 2015). In addition to allowing citizens to have a say in which issues should be addressed by local government, they provide tools for reading a city through instantly published reports that can be aggregated and analyzed. Given the complexity of infrastructure governance, one might wonder how such simple feedback mechanisms can be expected to accomplish so many things.

This chapter investigates how the different premises of infrastructure governance are enacted in the design of citizen feedback mechanisms and how this design influences users, who can include city agencies as well as residents. It looks at the assumptions embedded in the design of citizen feedback systems, assessing how these decisions shape the interaction between the citizen and the city. It makes a case for critically scrutinizing the political nature of these interfaces and argues that technical standards, protocols, and applications should be considered critical components for democratic discourse.

Analyzing reports submitted in the metropolitan area of Boston over four years between 2010 and 2014, this chapter compares the design aspects of two citizen feedback systems, CitizensConnect[1] and SeeClickFix. While the two systems are very similar in functionality and purpose, and their reported issues pertain to the same municipal departments, they have different histories: the former is an initiative by the city of Boston, while the latter is a private product. Because their only visible differences are the interface and the community of users, the two systems offer an ideal opportunity to study how interfaces and the system legibility they afford shape the interaction between citizens and local governments. The first part of the chapter focuses on questions of categorization—how different cities and their constituents frame urban problems.[2] The second part investigates the design paradigms and their assumptions about the user that guide design decisions for civic feedback systems.[3]

## CitizensConnect

Since 2008, Boston has operated what it describes as a constituent management system (CRM)—a database framework for managing complaints and requests submitted through the city hall's telephone hotline. A year later in 2009, based on an initiative by its office of New Urban Mechanics in collaboration with the mobile startup ConnectedBits, Boston was among the first cities to launch a smartphone application that allows people to submit reports directly from incident locations. Reports can be submitted anonymously and are referenced by the CRM with a case number allowing reporters to follow up on its resolution. Requests are made publicly accessible through the city's open data portal. CitizensConnect further supports the Open311 standard, launched in the same year as an interoperability initiative between different cities.[4] CitizensConnect represents a civic feedback system from the perspective of the government, enabling a direct connection to the municipal service providers. Through its simple and straightforward design, CitizensConnect has proved itself to be an exemplar of a successful mobile reporting system managed by a municipality.

## SeeClickFix

The second popular system that Bostonians can use, SeeClickFix, is a website and smartphone app released in 2007 and 2009, respectively (SeeClickFix 2007; Berkowitz 2009). Unlike NYC311 and CitizensConnect, SeeClickFix originated from a private social accountability initiative meant to improve communication between citizens and the local government. It was allegedly based on founder Ben Berkowitz's frustrations at getting graffiti removed from a neighboring building in his hometown, New Haven. In his words, "At first, we thought of calling it Little Brother, like 'Little Brother is Watching,' but then we realized we needed to be a bit more kind to government" (Roth 2009).

SeeClickFix is built around the idea of control from below. As in most accountability initiatives, collective action is a central instrument for creating public pressure. Citizens can create "watch areas" and assign them to public officials, who would then automatically receive all reports with or without their consent. In an ingenious second step, the startup then pursued service contracts with municipalities, offering to integrate SeeClickFix with the cities' internal operations and provide tools for managing and scheduling service requests (Harless 2013). Since 2011, SeeClickFix has been integrated with Boston's CRM, and in 2012, the Commonwealth of Massachusetts announced a collaboration under the name Commonwealth Connect with the company and Boston's Office of New Urban Mechanics (SeeClickFix 2011; State of Massachusetts 2012).

### What Exactly Is an Urban Problem?

What constitutes an urban problem beyond common nuisances like streetlight outages, potholes, or litter? CitizensConnect and SeeClickFix are just

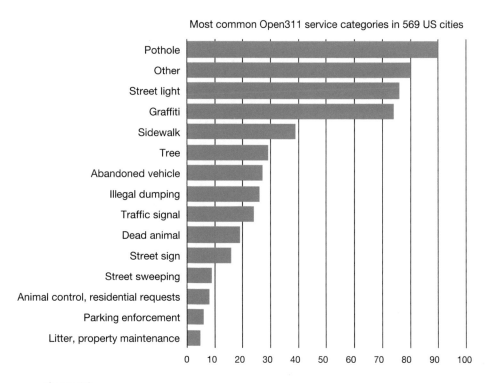

**Figure 6.1**
Most frequently used service categories in 569 U.S. cities using the Open311 standard, May 2015.

two examples of many similar projects connected through the Open311 standard, all of which answer this question differently. The question turns on subtle differences in terminology. The Open311 standard does not offer default categories for logging issues. Neither do commercial systems such as SeeClickFix or CitySourced. Boston uses "service requests" and "reports" interchangeably, while other cities draw a sharp distinction between the two. New York's NYC311 reporting system uses the word "complaints," while SeeClickFix speaks more neutrally of "issues." Baltimore provides separate categories for City Employee Praise and City Employee Complaint. Notably, only Bloomington, Indiana, offers a category explicitly called "suggestions." Comparing the service categories of over five hundred cities reveals a broad range of service categories, their specificity, and purpose (figure 6.1).

When launching a citizen reporting system, a city faces the task of routing the submitted reports to the correct recipient and dispatching work orders. In 311 call centers, an operator assigns the caller's concerns to an internal category. When using a digital interface, the citizen has to select a category from a predefined list of options. Ideally, the service categories listed in the interface reflect the city's own taxonomy. But this approach can quickly become unpractical and opaque for an outside user. At some point during 2014, the city of Toronto offered five service categories for reporting graffiti: Graffiti on a City road, Graffiti on a City bridge, Graffiti on private property, Graffiti on a City sidewalk, and Graffiti on a City litter bin. Daniel O'Brien of the Boston Area Research Initiative notes that the city internally defines a single broken streetlight as an "outage," but four broken streetlights in a row is a "large system failure."[5] While these distinctions help resolve departmental responsibilities, citizens read landmarks differently, use different concepts to describe issues, and have no insight into how each department categorizes its responsibilities. The physical affordances of the device also play a role. One hundred categories may be manageable on a desktop monitor but unusable on a smartphone touchscreen.

Based on technological changes and the feedback from users, cities continually refine the categorizations presented to users. Until 2013, the NYC311 desktop interface used a structured menu of roughly eighteen hundred different service categories. The smartphone app offered a similarly high number, forcing users to traverse a hierarchy of services to find a suitable category. A year later, the NYC311 app provided a much shorter list of twenty general categories. The desktop interface now guides the user through multiple steps, offering additional information and instruction.

Unlike the highly specific NYC311, which covers a broad range of services, CitizensConnect initially offered only three case types: potholes, graffiti, and streetlight problems. Since then, the Office for New Urban Mechanics has faced constant demand from users and city departments to expand the categories. Resisting the idea of overly specific categories, it introduced an *Other* category that covered any type of concern.

Just as external categories are constantly refined, internal categories are frequently modified, subject to what Geoffrey Bowker and Susan Leigh Star describe as the "practical politics of classifying and standardizing" (Bowker and Star 1999, 44). In this process, citizens instigate changes to the taxonomy. As one analyst involved in NYC311 recounted, citizen requests led to internal discussions such as how deep a pothole has to be to become the responsibility of the Department of Sanitation as opposed to the Department of Transportation.[6] By resolving citizen requests, the departments renegotiated their boundaries and relationships.

A similar reversal took place with Boston's CitizensConnect due to its simplicity and the mobility inherent in a smartphone app. Noticing that city employees often used the citizen app themselves, New Urban Mechanics created a version for city workers. Adapting the city's service categories for effective use in the field turned out to be a complex usability issue, requiring careful calibration of internal categories to the app's interface.

The question of standardizing the definitions of urban problems has frequently been discussed on the Open311 developer mailing list. Standardization would allow citizens to report issues regardless of administrative boundaries. For example, the greater Boston area includes the nearby municipalities of Brookline, Cambridge, and Somerville, which independently established their own reporting systems with their own apps, websites, and service categories. In a continuous metropolitan area, however, most residents are not aware of where one city ends and another begins. A program manager from Massachusetts noted on the Open311 mailing list:

The challenge is that every municipality thinks about each issue differently, including prioritization, sub-categories, who's responsible for service delivery, etc. ... While pretty much everyone deals with streetlight outages, potholes, and missing street signs, there is no common understanding of what each encompasses. Any vendor with a standard, fixed classification would be setting themselves up for a struggle to convince a municipality to abandon their existing classifications (no matter how informal). I'm not saying it couldn't be done, but it would be a struggle. (Heatherley 2012)

Citizens often have little patience for explanations that the issue does not fall under the city's responsibility because it is on private land, on a federal highway, or just outside the city's boundary. Cities mitigate these issues differently. Boston forwards outside requests to the relevant municipality or federal or state agency. The location-independent SeeClickFix app dynamically adjusts service categories and recipients based on user location. By offering geolocation tools and interfaces to third-party services like Twitter, software platforms are more malleable compared to the interfaces of physical infrastructures with baked-in standards. The consequence of this malleability is a constant renegotiation of interfaces between cities, city departments, companies, and individuals, including the occasional breakdown.

### The "Other" Issues—Implicit Themes in the General Category

As of early 2015, around one thousand cities in the United States accepted digital citizen reports. Larger cities tended to offer more service categories while small towns frequently used a single catch-all category. Most cities offered between three and twelve categories (including "Other") with the

**Figure 6.2**
Latent topics (selection) within the general "Other" category, Boston CitizensConnect, probabilistic topic models, figure by the author.

**Figure 6.2** (continued)

median at seven. In the case of Boston, the majority of reports submitted to CitizensConnect are categorized as Other. Is this a failure? Are the categories offered by the city inadequate for capturing the citizens' perceptions of issues? Or is this a desirable feature for keeping reporting informal?

One approach to investigating these questions is to examine whether the reports submitted in the general category contain salient, recurring themes that might as well be grouped into their own service category. Between September 2010 and August 2014, over forty thousand reports were submitted under the Other category via the CitizensConnect smartphone app in Boston. Methodologically, different approaches are possible to identify themes in large collections of unstructured text documents such as citizen reports. Grounded theory offers a systematic, iterative approach to developing conceptual models through qualitative comparative text analysis (Glaser and Strauss 1967). It would be, however, very difficult to manually analyze forty thousand documents in this way, and a small random sample might not be sufficient to account for variations over time and the relative differences in the saliency of the identified themes. To supplement my qualitative analysis based on the grounded theory approach, I decided to use an unsupervised machine learning technique called probabilistic topic models (Blei 2012). In this context, "topics" are lists of words inferred using latent semantic analysis (LSA) (Deerwester et al. 1990), which assumes that words appearing in close proximity across multiple documents indicate related meanings. Topic models are based on the assumption that collections of text documents contain multiple salient themes. For example, the archive of the *New York Times* might contain topics such as baseball, finance, or the conflict in the Middle East (Blei 2012).

In the context of probabilistic topic models, a "topic" is a collection of terms such as {red, traffic, accident, light, car}, which is likely to refer to reports about (potential) accidents and traffic lights. Meaning is expressed strictly in the relation between words, not in the terms themselves. For instance, the terms associated with *park* likely resolve the ambiguity in reference to a public garden or a stationary vehicle. As with any machine learning technique, the resulting topic models have to be taken with a grain of salt. Not every discovered topic represents a meaningful theme. Meaningful themes may be split across multiple topics, or a single topic can combine multiple unrelated meanings (Schmidt 2013).

The most frequently identified topic {parked, parking, cars, blocking, lane, car, fire, illegally} concerns issues of cars obstructing and blocking traffic. Example reports include "Bus blocking fire hydrant and street. Issue not resolved" or "Driver parked several feet off curb, obstructing traffic."

Another top-ranking topic concerned vehicles parking in residential areas without a permit. The second most popular topic {trash, garbage, street, sidewalk, left, bags, week, days} concerned garbage bags left on the sidewalk, for example, one citizen reported that "[redacted] has dumped many bags and loss (sic) bit of trash on street. This is not a trash pickup location and it is not trash day. This is a frequent issue with this address please cite them. Mission Hill is not a dump."

Other garbage-related topics include overflowing public waste bins and concerns about rodents ("neighbor across the street is continuously dumping rice and other food here, attracting rats, mice, and other pests"). Dangerous traffic situations are also frequently reported, as well as issues of overgrown weeds and fallen trees. Other salient themes include noise ("huge backyard student party at [redacted] VERY loud, underage?") as well as dog owners and homeless people. The topic {illegal, park, Segway, tour, city, gliders, plaza, hall} is an interesting case that captures a group of reports from a coordinated protest against a Segway tour operator: "Illegal Boston Gliders. Boston by Segway tour monopolizing Long Wharf pedestrian park."

How do the identified themes compare to the established categories? When run across all reports, not just those filed under "Other," the algorithm correctly identified themes that correspond to specific categories the respective reports are associated with. Analyzing the Other category separately found a high incidence of reports related to garbage and parking topics, which were overall reported more frequently than issues assigned to specific categories such as "damaged sign." Why does the city define a category for signs, but not one for parking or garbage? One answer might be that the city did not know better, and in fact, a trash category was added briefly following this analysis. However, there is also an important argument why the city would offer categories for relatively unpopular issues while excluding issues reported frequently. A *damaged sign* issue is fairly straightforward to resolve, while issues related to garbage require closer examination. Other matters such as traffic violations are not the responsibility of public works departments, and the city has good reason for not offering such categories.

## How Classification Shapes Interaction

How do the categories offered for use influence the kinds of issues reported? Categories do not just organize information; they also encourage particular types of reports and discourage others. As Susan Star and Geoffrey Bowker observe, "Each standard and each category valorizes some point of view and silences another" (Bowker and Star 1999, 5). Categories can nudge users toward submitting reports that are, in managerial parlance, more actionable. A complaint about a damaged sign can be added to a work queue immediately. Complaints that a park should look prettier or that the streets should be cleaner might be better suited for starting a broad conversation.

A few hints exist about how the choice of categories influences reporting. In September 2011, SeeClickFix announced a partnership with the city of Boston with the goal of integrating the SeeClickFix service into the city's CRM. As a result, the SeeClickFix interface adopted the same categories offered in CitizensConnect. Until then, SeeClickFix did not prescribe categories, and users could freely choose how to categorize the issue. Before the integration, graffiti was the subject of less than 1 percent of all submitted reports. After graffiti had become a category, the proportion of graffiti reports rose to approximately 4 percent of submitted reports, which is, however, still much lower than the 17 percent of graffiti reports reported to CitizensConnect (Offenhuber 2014). Notably, the city of Boston has operated a Graffiti Busters program with the support of volunteers since the late 1990s.

In designing categories for a feedback app, a city has to resolve the trade-off between using categories that reflect its internal operation and finding those that capture user perceptions. Going with the former lowers the friction of interpreting results and allows the city to be more responsive. Choosing the latter path can mean using no categorization or adopting categories created by users, known as folk-categorizations (Bowker and Star 1999, 59) or folksonomies (Voss 2007). However, the divide between the reports from citizens and public officials is less clear than it might seem. Many city employees use CitizensConnect in their daily inspections, and many citizens possess professional expertise in maintenance and repair issues.

The ambiguities inherent in defining categories are not necessarily a negative quality. They can encourage personal engagement and emphasize existing uncertainties (Gaver, Beaver, and Benford 2003). Commenting on their resistance to creating overly specific categories and incorporating all user suggestions, New Urban Mechanics noted that the "feature creep" of constantly adding new categories and functionalities ultimately diminishes a tool's usability.[7]

 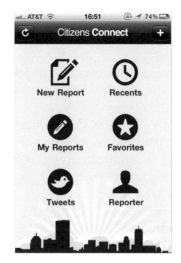

**Figure 6.3**
Screenshot of two smartphone civic-issue trackers used in Boston, both 2011 versions. Left: SeeClickFix (notice buttons "neighbors" and "my profile"), right: CitizensConnect.

## Design Paradigms of Feedback Systems

Categorization in a reporting system is just one means through which design influences interaction and data collection. Visual languages, system architectures, and the functionalities of the interfaces are equally important. Despite their different histories and goals, CitizensConnect and SeeClickFix have a remarkable number of similarities, which they share with the growing number of reporting platforms. Almost all systems offer smartphone apps with the same basic set of functionalities for submitting geocoded images and descriptive messages. All systems include mechanisms for tracking submitted requests and receiving responses. Frequently, users are able to browse other reports by time or location on a map. Despite these similarities, important differences remain that can be summarized in two different design paradigms that I will call the "direct route" and the "community-centered" approach.

The direct route model is characterized by restraint. The interfaces are limited to essential functions, and categories are fixed. The focus is on the submitted issue, not the reporter who remains anonymous. Communication channels are one to one between the user and a government recipient. Direct communication among citizens is rarely offered. With some exceptions, the direct route design is favored in city-developed systems focused on service delivery, with reporting categories closely corresponding to city services.

The community-centered model offers a rich palette of tools for many-to-many communication, aiming to foster a community of users who can evaluate and comment on other requests, even reopen issues that have been closed by the city. Users are encouraged to create self-descriptive profiles, often using pseudonyms. As in online forums, registered users are acknowledged for their contributions through a reputation system. Categories are less fixed, and users often have the opportunity to create their own or challenge existing categories.

The community-centered approach is more common in nonmunicipal systems focused on social accountability, such as SeeClickFix or the open source platforms Ushahidi and FixMyStreet (Okolloh 2009; King and Brown 2007). Lacking endorsement by the city, these approaches rely on an active community to attract participants and make the group voice stronger. A purpose secondary to reporting civic issues is to increase participation and encourage coordination through explicit and implicit channels. The many-to-many discussion resembles more a town hall meeting than a request for service.

Community-centered goals are not unique to volunteer-driven systems. Cities too are interested in encouraging participation and engaging citizens in infrastructure as a common good. However, their agenda is different, and ethical questions arise. According to Nigel Jacob from New Urban Mechanics, it is not appropriate for local governments to directly engage in building communities, but rather to listen to their concerns. Furthermore, when public services are involved, discussions are never open ended; there is always a filter of what issues are relevant in the context of urban maintenance. He frequently receives suggestions such as allowing people to vote on priorities, an idea the office resists.

Vibrant many-to-many conversations also present an interface problem. A mobile interface limits what can be accomplished through the size of the screen, the methods of user input, and the ability and willingness of users to learn complex interfaces. These constraints differ for private initiatives and municipalities that need to integrate their interfaces with an existing information infrastructure subject to legacy standards and historical contingencies.

The audiences for municipal and volunteer-driven systems are in many respects different. While social accountability initiatives seek to reach like-minded people who are willing to engage with a more complex system, municipalities need to maximize accessibility to reach the more casual user. Complex features that make an interface more expressive can come at the expense of accessibility. An IBM applications engineer advocated in an interview for a more minimalist approach, challenging the philosophy of rich social interfaces as the "the web way of thinking" that does not translate well into mobile applications and urban space.[8]

## Social Presence and Operational Transparency

Can interface design shape interactions? An important factor is how participants are represented. Earlier in the book, I discussed the concept of social presence, which refers to the capacity of a medium to convey verbal, nonverbal, and contextual information (Short, Williams, and Christie 1976). How a message is interpreted depends on the reputation of the speaker as much as on the words. Is the person a notorious complainer or promoting an agenda? In online communities, contributors are represented in terms of authority as much as by authorship. How frequently does a person contribute? Are his or her contributions appreciated by others? What are his or her areas of expertise?

Media scholar Judith Donath refers to the representations that combine self-description and a track record of activities as "data portraits" (Donath 2014, 187). She conceptualizes online communication in terms of nonverbal and implicit signals that can be implemented through gestures of acknowledgment or support rather than through explicit messages. Online representations of oneself can also be a powerful motivator when reporters see their issues being acted upon and receive feedback from the community.

The differences between the design of the two reporting interfaces are striking in terms of how participants are represented. CitizensConnect offers no direct communication among users. A city department can respond to a user request with a standard reply or a customized message. Not only does SeeClickFix offer direct interaction among participants, but it also represents citizens and city officials the same way, drawing no principal distinctions between them. The Open311 standard supports little social presence and is strictly limited to one-to-one communication between a submitter and a department. To circumvent this limitation, systems such as CitizensConnect post reports on Twitter, enabling more channels of interaction among users.

Beyond the representations of users and governors, operational transparency involves the legibility of the city's actions and priorities, for example, by showing where its workforce is currently active (Buell, Porter, and Norton

2014). The city's website fulfills this function to some extent, but examples where users can watch maintenance transpire in real time are rare. The city of Boston has experimented with action shots that show in their mobile app how a submitted issue is being fixed and currently shows a picture of the respective city unit next to resolved requests. During the 2015 snowstorms, the city created a temporary website[9] showing the real-time location of all municipal snowplows. It can be hypothesized that cross-indexing the locations of municipal work crews on a map of submitted reports conveys a more realistic image of what is involved in urban maintenance than seeing an issue isolated in the context of an overall task list.

Besides the elements that increase the visibility of users and issues, some elements limit visibility. All feedback platforms must deal with reports that misuse public visibility and anonymous reporting for the purposes of advertising, data collection, harassment, or vandalization through a flood of unrelated, often automatically generated reports. The boundaries between moderation and censorship can be difficult to draw. However, a city usually faces a more basic problem with a lack of resources to review submissions. It therefore has to rely on platform design to manage report visibility and add friction to the submission process. Submitting a report has to be convenient enough to encourage citizen participation while being inconvenient enough to deter spamming and other forms of abuse.

Even if reports are published immediately without review, they can be made more or less visible. For SeeClickFix, the website plays an important role; it shows all reports in the context of similar issues reported by others, and the responses from officials and other users. The textual descriptions, which often take a critical tone toward the city, therefore have a high degree of public visibility. In the municipal system, this is less so. As of this writing, the home page for the city of Boston prominently features its 311 system, but the page for browsing the submitted requests is no longer directly accessible through the city's home page. In previous versions of the home page, the reports page was better integrated, but the visitor still had to traverse a series of links in order to read the submitted requests. While the city's open data portal offers access to the real-time data set of 311 service requests comprising more than thirty data and metadata columns, the column containing the text of the actual complaints submitted by citizens is missing.[10] Users of the mobile app can still read reports on their device, and technically savvy users can still access the text descriptions through the Open311 API. Overall, however, the submitted requests have become less visible to the public over time.

Unrestricted public accessibility in itself, however, does not ensure high visibility, which can be demonstrated through the website for browsing reports submitted to the city of Boston.[11] The site uses a Twitter-like interface that displays a real-time stream of reports as they come in. The display is public but ephemeral: it offers little assistance to search for a specific report or to compare reports from different times. With about a hundred reports submitted daily, it becomes difficult to locate a specific report after a few months through this particular interface. One could call such a design principle "opacity through transparency"; as all reports are immediately published, information is obfuscated precisely because of—not despite—the amount of information.

These design decisions may introduce enough friction to discourage spammers and vandals, but the limited visibility also has implications for critics of the government's priorities and decisions who use the system to voice their concerns. Every design decision, however accidental or whatever the underlying intent, has consequences for the politics of visibility.

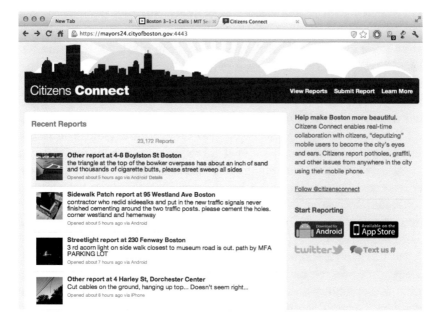

**Figure 6.4**
Screenshot of the CitizensConnect website used to browse reports in its version from 2010.

## Effects of the Interface on Submitted Reports

To what extent are these considerations reflected in user submissions? I addressed this question through a comparative text analysis of a sample of two thousand reports submitted to SeeClickFix and CitizensConnect (Offenhuber 2015). Again, Boston is a suitable case study because it has used the popular CitizensConnect for a number of years and has also integrated its services with SeeClickFix. With only the interfaces differing, it is possible to investigate the effects of design decisions on submitted reports, which were largely consistent across both interfaces, with a few exceptions. Comparing reports submitted to both systems shows that infrastructure repair issues are more prominent in SeeClickFix, while issues concerning graffiti and litter are notably absent. In CitizensConnect, these issues account for more than a third of all reports (figure 6.5).

Although most reports are written in a neutral and factual tone on both systems, SeeClickFix reports tend to be more critical, meaning that they emphasize the importance of the issue and urge the city to act: "3rd report of crumbling stairway. Getting very dangerous" or "Light goes out periodically then comes on slowly. Dangerous area for drugs, assaults. Please fix. Thanks." Very critical reports involving blame and shame occur in about 5 percent of reported cases in both systems. For example: "Paint the white lines. It's horrible that the lines have been missing here for over 1 year. You are on notice, if someone gets hurt the city is liable. Shame that there is a school 20 feet away..." Not just the city is shamed; frequently, fellow citizens are as well: "Our neighbor always brings her daughter and dogs to poop in front of our house and they live in 433 in the 1st and 2nd apartment. I called the Animal Control for 2 years and nothing changed." SeeClickFix has a smaller proportion of reports that directly accuse other citizens. Irate reports in CitizensConnect are often triggered by trash and litter issues, which do not play a major role in SeeClickFix. Conversely, infrastructure repair issues frequently trigger critical reports on SeeClickFix, while such matters are among the most neutrally discussed on CitizensConnect.

The observed differences are consistent with the more private, service-oriented, one-to-one nature of CitizensConnect versus the more public, social accountability-centered, and discursive nature of SeeClickFix. Users might hesitate to report personal grievances such as graffiti, litter, and traffic violations to a publicly visible forum. A review of the SeeClickFix website shows that this concern is not baseless. SeeClickFix users are united in opinions about larger infrastructural issues, but when private issues emerge, so do multiple controversies. A report about a "stolen" parking

| Incident type | CitizensConnect N | % | Delta % | SeeClickFix % | N |
|---|---|---|---|---|---|
| Graffiti | 210 | 18% | | 2% | 16 |
| Trash/litter | 211 | 18% | | 4% | 29 |
| Ice | 37 | 3% | | 0% | 3 |
| Plants | 42 | 4% | | 2% | 11 |
| Other violation | 40 | 3% | | 2% | 13 |
| Animals | 16 | 1% | | 1% | 7 |
| Test/unknown | 13 | 1% | | 1% | 8 |
| Traffic | 64 | 5% | | 6% | 39 |
| Social issues | 7 | 1% | | 1% | 10 |
| Infrastructure improvement | 39 | 3% | | 13% | 92 |
| Infrastructure repair | 493 | 42% | | 67% | 466 |
| **Total** | **1172** | **100%** | | **100%** | **694** |

| Motivations, concerns | N | % | Delta % | % | N |
|---|---|---|---|---|---|
| None specified | 345 | 29% | | 17% | 118 |
| Aesthetic concerns | 174 | 15% | | 5% | 33 |
| Other people's behavior | 108 | 9% | | 7% | 47 |
| Public health/sanitation | 37 | 3% | | 2% | 11 |
| Dissatisfied with the city service | 38 | 3% | | 2% | 12 |
| Bad personal experience | 15 | 1% | | 5% | 32 |
| Safety concerns | 111 | 9% | | 17% | 116 |
| Submitting ideas | 39 | 3% | | 11% | 76 |
| Concerns with disrepair | 305 | 26% | | 36% | 249 |
| **Total** | **1172** | **100%** | | **100%** | **694** |

| Tone of report | N | % | Delta % | % | N |
|---|---|---|---|---|---|
| No text | 234 | 20% | | 15% | 104 |
| Neutral | 577 | 49% | | 46% | 317 |
| Plea | 64 | 5% | | 3% | 20 |
| Very critical | 38 | 3% | | 3% | 19 |
| Friendly | 54 | 5% | | 4% | 30 |
| Critical | 205 | 17% | | 29% | 204 |
| **Total** | **1172** | **100%** | | **100%** | **694** |

| Themes | N | % | Delta % | % | N |
|---|---|---|---|---|---|
| Reporting a specific person | 118 | 10% | | 4% | 26 |
| Demanding accountability | 60 | 5% | | 4% | 30 |
| Strong language | 37 | 3% | | 3% | 21 |
| Suggesting improvements | 53 | 5% | | 12% | 81 |
| Concerns for safety | 143 | 12% | | 20% | 140 |

**Figure 6.5**
Relative differences between reports submitted by CitizensConnect (CC) and SeeClickFix (SCF) users based on a randomly drawn sample of reports.

spot quickly turned into a broad discussion about social norms for these types of situations.

Users on the two platforms tend to explain issues differently. To justify urgency, a larger proportion of SeeClickFix users invoke public safety: "This is a terrible intersection. Constant beeping every 5 mins disturbs the neighborhood. I'm afraid there will be an accident here all the time. I've almost been hit several times."

Again, the more public nature of reports displayed in SeeClickFix and the desire to mobilize other users are a possible explanation. But not all reports were critical of the city's services. Frequently, reporters offered ideas and suggestions regarding how to resolve a specific situation. For example: "Google maps says this area is a park. Doesn't look like a park to me. This area has one of the best water views in Boston and looks awful. There should be a park bench or something nice there. Also the guardrail is very old looking and beat up. Makes the neighborhood look disgusting. The whole area is very un-looked after."

Or: "Fallon (sic) field playground climber has come undone. Requires big-ass tamper-proof Torx bits. I think that's all that's needed."

Many issues are of a similarly technical nature, but sometimes social tensions become apparent in the reports: "PANHANDLER/BEGGAR ..., holding door open (to tracks 1 and 3), implying he's asking for money. I shouldn't have to put up with this while I'm paying $235 a month for my commute. Please have him removed and reinforce he should seek assistance elsewhere."

In about 5 percent of reports in both systems, accountability is demanded: "Whoever got paid to close this report ripped off the taxpayers TWICE." A reporter in East Boston complained about unequal service provision, writing, "Does one have to live in a posh neighborhood to get something done? Isn't an abandoned U-Haul truck a security concern?"

Despite its integration with city services, SeeClickFix presents itself as relatively independent. It therefore receives a higher percentage of critical reports, and its service requests tend to be less straightforward. The higher public visibility and lower expectation of privacy likely contribute to the different style of reports, which are more open ended and frequently emphasize the public good and safety implications.

### Prioritization

Any feedback system is only as good as the city's response at fixing the problem. A feeling that the city's capacity to respond to requests does not match the convenience of the tool leaves users frustrated: "72 days ago I posted this under case id 101000405068 city forward info and details to DCR and forgot about it 72 days later nobody even care (sic) about this. What is the purpose of this citizens connect if we voters are not taken in consideration by just simply being ignored ..."

Not surprisingly, issues such as actual infrastructure damages and violations receive a faster response than reports that raise open-ended, more diffuse issues such as suggestions for infrastructure improvements and discussions of civic issues (table 6.1). However, the tone of the report makes a difference. Reports using highly critical language were resolved quickest—in other words, the squeaky wheel gets the grease. Response time is, of course, not an appropriate measure of service quality, but it does indicate the priorities of the city or, better, the city's perception of citizen priorities.

Beyond questions of response time, both the city of Boston and SeeClickFix frequently emphasize the value of citizen feedback systems to facilitate coordination and self-service. Late Boston Mayor Tom Menino frequently cited the following exchange between two citizens over the CitizensConnect platform. A report from February 2011 reads: "Possum in my trash can. Can't tell if it's dead. Barrel in back of 168 west 9th. How do I get this removed?" Before the city's animal control was able to respond, neighbor Susan Landibar submitted another report: "Walked over to West Ninth Street. It's about three blocks from my house. Locate trash can behind house. Possum? Check. Living? Yep. Turned the trash can on its side. Walked home. Good night, sweet possum" (Gaffin 2011). However, such interactions between citizens are not directly supported by the interface, which does not allow citizens to comment on each other's questions without submitting a separate report. Through its more community-oriented interface, SeeClickFix actively encourages such coordination among citizens when cases fall outside the city's responsibility. During a Boston snowstorm, a neighbor initiative used the platform to organize the snow removal from private cars and driveways within the community (Snowcrew 2012).

## Conclusion: The Designer as Regulator

Open data portals and real-time information feeds do not mean that a city or a provider has no control over how these data are perceived by the public. As these case studies from Boston make clear, digital interfaces do not merely augment public discourse; they produce and increasingly regulate it. The design of feedback systems determines the visibility of the reported

**Table 6.1**
Table 6.1 Citizens Connect: average response time by the city in days for closed issues (N=849)

| By service category | Days open mean | median | By incident type | Days open mean | median |
|---|---|---|---|---|---|
| Graffiti | 17.3 | 7.3 | Plants | 37.9 | 2 |
| Other | 16.8 | 1.5 | Social issues | 35 | 1.8 |
| Streetlight | 13 | 3.8 | Infrastructure improvement | 19.8 | 8.2 |
| Damaged sign | 12.4 | 8.7 | Graffiti | 17.7 | 7.3 |
| Pothole | 3.1 | 1.3 | Infrastructure repair | 15.7 | 2.1 |
| Sidewalk patch | 2.9 | 0.9 | Other violation | 11.1 | 1.2 |
| Unshoveled sidewalk | 2.6 | 2.8 | Trash / litter | 2.5 | 1 |
| Roadway Plowing/sanding | 1.5 | 1 | Ice | 2.4 | 2.3 |
|  |  |  | Traffic | 2.4 | 0.9 |
|  |  |  | Test / unknown | 0.7 | 0.7 |
|  |  |  | Animals | 0.6 | 0.6 |

| By motivations expressed | mean | median | By tone of report | mean | median |
|---|---|---|---|---|---|
| Ideas / discussion civic issues | 31.2 | 6.6 | Friendly | 20.8 | 3.2 |
| Safety concerns | 22.2 | 3.1 | Plea | 16.2 | 1.5 |
| Concerns with disrepair | 18.7 | 2 | Neutral | 14.3 | 1.9 |
| Bad personal experience | 13.7 | 1.1 | Critical | 13.6 | 2 |
| None specified | 11.4 | 3.7 | Very critical | 10.1 | 0.8 |
| Aesthetic concerns | 5.1 | 0.9 | Report contained no text | 7.4 | 2.8 |
| Issue with other people's behavior | 3.7 | 1.2 |  |  |  |
| Dissatisfied with the city service | 3.4 | 0.8 |  |  |  |
| Public health / sanitary concerns | 1.9 | 1.3 |  |  |  |

issues, the people submitting and discussing them, and the response and actions taken by the city. The language and categories used in the interfaces frame what can be reported, and the modes of communication available govern the interaction among the actors. Through the design of the interfaces, interactions can be encouraged or discouraged, as well as steered toward a specific issue or an open-ended discussion.

It may be accurate that, despite the angry tone of some reports, discussions about snow removal and garbage on the street are business as usual for the city and hardly controversial in the larger picture of infrastructure governance. However, there is no reason to assume that a city struggling with the impact of a massive snowstorm might not be tempted to remove a pile of embarrassing complaints about insufficient snow removal. The idea of local government as an open-source "urban operating system," sketched out by Government 2.0 advocates can be misleading, since even a technically mediated system is still negotiated by human agency. Compared to the idea of an urban operating system governed by incorruptible algorithms, design gestures are not subordinate to a totalitarian algorithmic logic, but instead are products of countless human decisions and negotiations that do not necessarily follow a comprehensive scheme. Often, design decisions are accidental, implemented by different people without coordination and without awareness of their implications. Components might be unintentionally broken by upgrades and consequently removed. By focusing our attention to design instead of the abstract logic of the algorithm, we can foreground governance as an ongoing conversation and negotiation.

## The Politics of Interface Design

While interface designers stress the importance of responding to user needs, the opposite is often true. The interface configures the user, structuring her or his behavior according to the intentions of the designer and the constraints of technology (Woolgar 1991). Design choices can guide a conversation between a citizen and a city official toward either open-ended deliberation or efficient problem solving. By regulating and framing the interaction between citizens and the city, they have consequences for the governance of an infrastructure, assuming a role that is deeply political.

Although designers object to the view that interface design is a cosmetic task, they are frequently unfamiliar with their role as a political mediator. Often unknowingly, interface designers find themselves responsible for regulating and governing behavior. Not all submissions are desired by the city, including spam, personal attacks, or reports that touch on matters that officials prefer not to discuss. Design offers a way to manage the torrent of submissions and nudge it in a particular direction.

Using the example of air travel, geographers Robert Kitchin and Martin Dodge describe how the governance of the physical space is contingent upon interactions mediated through digital interfaces that are increasingly exposed to the user through ticket booking, check-in, and recently self-service passport control (Kitchin and Dodge 2011). In these cases, interface design assumes a crucial role in guiding, informing, and shaping data entry. As this chapter has demonstrated, the political implications of interface design for infrastructure governance are present at multiple levels or registers:

• Terminology and categorization, whether institution-centered or user-centered, general or specific. Comparing the categorizations used in different cities reflects not only local characteristics and issues, but also different philosophies of citizen-city interaction.
• How digital interfaces constrain and facilitate communication on the input side and how they govern the public visibility of contribution on the output side. As a consequence, how these arrangements encourage or discourage certain expressions and behaviors.
• The seamful and seamless aspects of a system—can an amateur or expert appropriate and extend the system? Which visual aspects serve the novice user to get a better sense of what happens inside the black box of the system, even if not all technical details are accessible? Conversely, which aspects of the system should remain invisible to prevent obfuscating crucial information?
• Specifications of standards and protocols such as Open311. Protocols and open standards should be considered as part of the democratic discourse as well as the basis of open source software ecologies.

**The Digital and the Human Interface**
The interfaces of both SeeClickFix and CitizensConnect are not stable in time, but subject to constant evolution. While working on the case studies, the user interfaces, APIs, and the underlying data structures have changed multiple times, making data collection challenging. SeeClickFix at some point abandoned its open-ended categories and adopted Boston's service types, facilitating better integration with the city. The city of Boston has continuously iterated the design of websites and apps that display of citizen reports. These design decisions have consequences for the visibility of reports. The submitted texts of reports were initially accessible through the

site, which is no longer the case. In return, the city has expanded the scope of its data API, making more sophisticated queries possible. At any given time, both services have offered a suite of different protocols, file formats, and interfaces to access and work with data, but this multiplicity has also changed over time.

While no universal interface can exist that is equally accessible to all users, a multiplicity of channels (phone, letters, emails, tweets, and so on), interfaces (websites, apps), and standards (Open311, SeeClickFix's own API) can exist in parallel, offering different alternatives for access.

No discussion of digital governance can avoid the digital divide. Not everyone owns a smartphone or has Internet access. Nor is everyone comfortable with or capable of using digital interfaces to access government services. But at the same time, this divide is not binary—access or no access to digital services—but instead manifests itself in a more nuanced manner, in different expectations and attitudes toward services. The squeaky wheel fallacy assumes that the absence of negative feedback means that there is no problem, which results in the concentration of public resources on the most vocal neighborhoods, creating a system that captures wants rather than needs.

Access remains a challenge, and the city of Boston responds by offering many different channels of access rather than prioritizing one channel. The multilingual version of Boston's CitizensConnect feedback system used simple SMS text messages, but it was not successful due to the cumbersome user experience. Inspired by popular food truck services, City Hall to Go brings government services to remote neighborhoods by truck (New Urban Mechanics 2014). According to Nigel Jacob and Chris Osgood, City Hall to Go reflects New Urban Mechanics' goal of providing a space for civic experiments without relying on technological mediation as a universal solution. As Jacob noted, "When people come to us with ideas, there is no interface in-between; they are already inside."[12]

Through the design of different feedback systems, cities are walking the line between managing criticism and inspiring engagement. It might seem counterintuitive that city departments would enthusiastically embrace social accountability mechanisms that can potentially put them under public pressure. But as the example of GuttenPlag has shown, if governments do not actively support building these systems, citizens might do so anyway, with or without their approval.

# Epilogue to Part III: Tool or Therapy? Critiques of Civic Technologies

Citizen participation is almost without exception touted as a positive value that seems to unite otherwise incompatible ideologies. For the left, participation is a way to empower the weak; for the right, it is a way to emphasize individual responsibility. Governments and the nonprofit sector embrace participatory platforms because they promote the public good; companies, because they create new markets and opportunities. Across all these diverse perspectives, it is not always clear whether participation is seen as a means to an end or a value in itself. Part III of this book examined the interactions between citizen and government through civic technologies at the granular level, but has not addressed the role of participation from a broader perspective. Some questions remain about the purposes and politics of civic technologies.

As policy researcher Sherry Arnstein noted, perhaps somewhat sarcastically, "The idea of citizen participation is a little like eating spinach: no one is against it in principle because it is good for you" (Arnstein 1969). Unless, as she argues, participation raises questions of power distribution, in which case it quickly becomes a controversial issue. Arguing for a more nuanced view of participation, she presents a ladder of citizen participation as a typology of involvement ordered by the degree of control afforded to the participant—ranging from pseudo-participation that placates users to real citizen control with delegated power.

It makes sense to use a similar lens for examining the scope of participation in civic technologies, considering the generous use of the term in this field. As noted earlier, participating in service provision is a somewhat ambivalent action. It is often less an enactment of democratic values and more a coping strategy for living with poorly functioning services. It can be a burden as well as a value. Since urban services are participatory almost by definition, it is imperative to take a closer look at the functions of participation and its beneficiaries.

## Participation as Compliance

On the lowest rung of the ladder, participation can simply mean using a system as intended by its designers. In this sense, urban services are participatory by definition; Arnstein would probably describe it as nonparticipation. Waste management service providers and recyclers, for whom configuring the behavior of the user in accordance with the system is often an operational necessity, nevertheless employ the rhetoric of participation.

Besides familiar calls to participation such as educational programs, appeals to civic duty, or slogans such as Don't Litter, Recycle More, and Keep America Beautiful, compliance is increasingly manufactured through design techniques known as gamification or nudging. For example, a smartphone app from aluminum producer Alcoa rewards users with points for every can returned to a recycling center (Alcoa 2010).

"Gamification," which is "the use of game design elements in nongame contexts" (Deterding et al. 2011), involves, for example, rewarding desirable behaviors with points and encouraging users to compete against each other and compare their scores. "Nudging" is a related approach to influence behavior without enforcement by carefully tweaking the choice architecture of the environment in which citizens make decisions (Thaler and Sunstein 2008). Nudging starts with acknowledging the behavior-shaping role of default settings offered to the user, but can also employ more aggressive forms of behaviorist-style self-conditioning.

Both gamification and nudging are seemingly apolitical, outcome oriented, and free of moral appeals or coercion. Nevertheless, one must still question who defines what constitutes desirable, positive behavior. What does it mean when an aluminum producer rewards citizens for every single beverage can they recycle? While the app might be the result of a genuine concern for the environment, it also offers incentives for buying more cans in the first place and reminds users about their environmental responsibilities. In the words of political scientist Michael Maniates, this form of individualization "understands environmental degradation as the product of *individual* shortcomings ... best countered by action that is staunchly *individual* and typically *consumer-based* ... It embraces the notion that knotty issues of consumption, consumerism, power, and responsibility can be resolved neatly and cleanly through enlightened, uncoordinated consumer choice" (italics in original; Maniates 2001, 31).

## Participation as Feedback

The next level of user agency is feedback. Users can either be a passive source of information about their needs and wants or actively submit information. As a passive source, users may not be aware that they produce data simply by using a service. Participation in polls and citizen report cards such as New York's Project Scorecard, which has been deployed since the 1970s to gauge street cleanliness (Melosi 2004, 252), requires more involvement but is still only one-way communication. In more active forms of feedback, users volunteer information by submitting complaints or service requests.

Ideally, feedback mechanisms benefit the provider, who can gather performance metrics that allow for a more targeted service provision, as well as the users, as long as their concerns are addressed and acted upon. In practice, this has not always been the case. In a comparative study of feedback mechanisms used in development projects, many initiatives failed to improve service (Cavill and Sohail 2004). Despite the simplicity and immediacy of the short route between users and private service providers, users felt that their complaints were ignored. The long route of accountability, for example, by approaching elected public officials with complaints about service provision, yielded better results.

Another concern is that feedback systems reinforce existing inequalities, in which those with the loudest voice and biggest influence receive the most attention and resources. Studies of citizen feedback systems have shown that residents in deprived areas often complain less compared to better-serviced residential areas of the middle class (Verplanke et al. 2010; Martínez, Pfeffer, and van Dijk 2009).

## Participation as Oversight

If citizen feedback is connected to enforcement mechanisms, participants gain a stronger voice, which can also benefit the systems governors in several ways. Cities can enlist constituents to monitor private service contractors such as waste haulers or recyclers. In many cities, citizen satisfaction is used for evaluating contractual performance. In international development, lenders and agencies may attempt to prevent waste and corruption by calling on communities to monitor the use of funds for the construction of streets or sanitation systems. This practice of participatory monitoring also entails involving a community in measuring the quality of service provision (Estrella and Gaventa 1998). When accountability mechanisms are absent, citizens might form watchdog initiatives themselves, forcing local governments to listen to them using informal means of public shaming.

Integrating citizen data into service provisions like trash collection faces a number of challenges, including the credibility of information generated by amateurs. The same, however, is true of official sources; data collected by public officials and professionals often are equally prone to biases. A study of street maintenance in New York City found that citizen-generated data are in some cases more accurate than authoritative data sources, inde- pendent of socioeconomic factors (Van Ryzin, Immerwahr, and Altman 2008).

Increased responsibilities for users and higher expectations of data qual- ity require a better understanding of who the users are. Different groups, ranging from the casual contributor to the fully committed expert, have different motivations to contribute, which need to be addressed in the architecture and design of the system (Coleman, Georgiadou, and Labonte 2009).

Structural constraints of voluntary data collection are harder to address. Volunteer data are nonprobabilistic and subject to various systematic biases. Systems such as Wikipedia generally have a highly asymmetrical relation- ship between readers and contributors: 2.5 percent of the users generate 80 percent of the total content (Rafaeli and Ariel 2008). Jacob Nielsen coined the 90–9-1 rule to reflect that 90 percent of social media users consume but do not contribute, 9 percent contribute occasionally, and only 1 percent contribute on a regular basis (Nielsen 2006). Similar distributions are found in many other volunteer-driven systems. In the context of infrastructure management, these sampling issues limit the usefulness of generated data, highlighting a conflict between the expectation of a homogeneous and reli- able service and the uneven nature of user participation. Cities address this issue by targeting a smaller group of motivated expert users rather than trying to overcome the structural limits of participation.

**Participation as Co-governance and Self-organization**

Within the conventional paradigm of urban services, involving users directly in service provision is the exception rather than the rule. The orga- nization and maintenance of self-organized infrastructures are not trivial; building an infrastructure is easier than maintaining it. Nevertheless, a number of examples exist. Nonprofit organizations frequently maintain physical infrastructures based on volunteer work. The Appalachian Trail Conservancy in New England, for instance, maintains an extensive network of hiking trails entirely through the labor of over six thousand volunteers. Other examples include resident-driven paper and cardboard collection systems in the Netherlands (De Jong and Mulder 2012).

Such models of community-based service provision can be successful if the volunteers are at the same time the beneficiaries of the service. Co-production, involving residents and recipients in the planning of infrastructure projects, can be successful if it takes the needs and knowledge of locals into account (Ostrom 1996; Ibem 2009, 130).

Volunteerism in service provision is frequently accompanied by a rhetoric of empowerment, which in the hands of a power holder can quickly become condescending. Participation can be a burden. Paraphrasing policy expert Peter Schübeler, why should citizens in poorly serviced neighborhoods concern themselves with service provision when the local government provides better service to neighborhoods of a higher socioeconomic status (Schubeler 1996, 32)?

The failure of the Big Society policy initiatives under former British Prime Minister David Cameron illustrates that volunteer-driven initiatives rarely work as a replacement for public services, but on the contrary, require public support to flourish. After the initial success of the volunteer program during the 2012 London Olympics, the initiative has received criticism for masking an agenda of dismantling public services and using volunteers to compensate for it. Studies suggest that volunteerism declines when government intervention decreases (Bartels, Cozzi, and Mantovan 2013).

### Criticisms of Participatory Models

The most fundamental critiques of participatory models of infrastructure provision and maintenance concern the concept of participation itself. The value of participation is often taken for granted as a means as well as a prerequisite for a just and inclusive society, to the point where participation has been called the new "grand narrative" (Cooke and Kothari 2001, 139).

When engagement is celebrated for the sake of engagement, the IKEAization of service provision can create an atmosphere of activity that can distract from larger systemic challenges. As waste management expert and environmental scholar Samantha MacBride observes, Busy-ness is a handy method of maintaining the status quo yet is simultaneously active, optimistic, and often makes people feel better (MacBride 2012, 6). It is desirable for citizens to be able to report potholes or broken electricity poles. It is even better if the city responds to these requests in a timely manner. However, the energy and cost that go into incremental fixes can come at the expense of more comprehensive solutions, such as using a road surface less prone to potholes or burying power lines to minimize outages.

The virtues of incrementalism and solution-oriented attitudes can lead to stagnation by limiting a society's gaze to the inconveniences of the everyday. These unintended effects show that participation for its own sake is not enough if principal benefits are simply assumed to exist. One must look closer at the model through which participation is enacted.

# Conclusion: A Case for Accountability-Oriented Design

For better or worse, we have entered a new paradigm of public service provision characterized by more user involvement and a blurring of the boundary between providers and users. This paradigm affects urban infrastructures and public services, including the four big utilities of water, electricity, sanitation, and transportation. Terms such as "(prod)user," "expert amateur," and "civic hacker" indicate that users have to some extent become service providers themselves, turning their cars into taxis and their bedrooms into hotels, generating electricity and monitoring public service quality, reporting issues through online tools or challenging official data using their own sensor networks. Individuals book their own flights, execute their own money transfers, choose between different service providers, and leave detailed traces of their choices and preferences in the process. Not all aspects of this *infrastructural inversion* are voluntary or make the interactions with and within urban systems less burdensome. The increased level of participation in urban services also enables new accountability instruments and tools for mobilization, which promise the public more voice in deciding matters of infrastructure governance.

Unlike earlier phases of decentralization promoted by market-oriented public management reformers, a diverse group of actors is driving this current infrastructural inversion, each with different goals and politics. Local governments, NGOs, technology startups, and data activists use social media technologies to provide, augment, and scrutinize public services. The similarity of their tools and their coordination and mobilization tactics, however, can distract from ideological differences and diverging visions of what constitutes "good governance." Depending on perspective, civic technologies may appear neoliberal or neo-Weberian, critical or service-oriented, tools for deliberation or technocratic governance. An application that helps neighbors fix infrastructural issues among themselves may be designed to foster shared civic values by appropriating the transactional mechanisms used by commercial services such as Uber or Airbnb. A city could use this app to make public service provision more useful and meaningful, or offer it as an excuse to withdraw from service provision and shift the responsibility to the public.

The described paradigm is information-centric: based on the assumption that all urban issues are problems that can be, in one way or another, addressed by exchanging information and improving coordination. This involves assessing needs and issues either directly by collecting feedback or indirectly by appropriating available data sources that can be used as proxies. It involves making service provision more targeted by intervening only where service is needed and measuring outcomes based on available data and the metrics of choice. Civic technologies promise to open new spaces for deliberation and help create public pressure by coordinating social action on a massive scale when services fail. Yet this information-centrism is problematic in several ways. Information does not automatically translate into action, and data generated by sensors or volunteers offer an image that is necessarily incomplete and shaped by particular interests. As acknowledged even by optimistic voices of the development sector, information technology is no "equalizer" that mitigates economic and social differences (World Bank 2016). Working with public records can resemble searching for a needle in a haystack and, paradoxically, transparency and open data initiatives can obfuscate relevant information by creating bigger haystacks.

Civic technologies are not just a pragmatic means for simplifying communication with cities or collecting information about issues of concern. They create a world of informational objects through which users describe and perceive their environment. This world, in Woolgar's terms, configures its users, requires specific knowledge, and favors certain behaviors and forms of expression (Woolgar 1991). Interface design determines how users are represented and shapes their interactions. By simplifying communication and shortening distance, civic technologies informalize the interactions between citizens and governments. At the same time, it makes these interactions more formal by recording a persistent and identifiable trace that is not bound to the specific context of the interaction, one that can be aggregated and analyzed.

As they define the objects, rules, and governance of this interface world, interface designers are often not aware of their implicit ontological claims. They have, however, the capacity to make infrastructures legible in a way that acknowledges their heterogeneous natures. This can involve emphasizing the seams between system components, providing clues that indicate activity, creating interfaces that are socially translucent rather than transparent, developing visual languages to express the processes of governance, or supplying references about where to get additional information. At the same time, their representations of socio-technical systems are always partial and incomplete, based on perspectives that are never universal. Legibility always serves a specific purpose.

## The Dataization of the Waste System

Perhaps more than other infrastructures, waste systems have so far resisted pervasive *dataization*. Compared to some of the more esoteric issues of smart cities and data-driven urban management, waste management struggles with fundamental data issues that affect policy decisions. While some aspects of waste management are increasingly captured—examples include the material composition received and processed by automated MRFs, collection volumes at the household scale with the help of RFID tags, and remote sensing methods to identify informal dumpsites (Hannan et al. 2015)—these data sources remain islands of information in a largely opaque system. These islands may be limited to a specific purpose, such as the economics of collecting and valorizing specific recyclables including bottles, metals, papers, and plastics. They may be limited to a specific area by, for instance, the idiosyncratic and incompatible taxonomies and data collection methods in different states, or more importantly, by the specific challenges faced by developing and developed countries (Wilson 2007). Connecting these islands is complicated by the fact that existing data are often based on unknown or incompatible methodologies, starting with widely diverging definitions of such fundamental concepts as municipal solid waste.

These issues make the waste system a good case study for investigating the mechanisms and limitations of monitoring practices for socio-technical systems, especially considering the political and contested nature of the waste system's definitions and monitoring procedures. The case studies discussed in this book investigate the waste system from the bottom-up perspective, investigating the movements of waste across state boundaries, the informal organization of collection, and the processes of urban maintenance through citizen reports that cover waste and sanitation issues. Constructing an image based on the data from these studies required engaging with technologies of geolocalization, open data repositories, data from other participatory initiatives, and regulatory databases.

In conceptualizing infrastructure legibility, I have compared notions of urban legibility described by James Scott and Kevin Lynch. Scottian legibility utilizes a perspective from above and involves creating system-wide, simplified representations based on standardized symbolic conventions. Lynchian legibility, on the other hand, means constructing an image of the system from below, based on a heterogeneous set of clues and traces. While Scottian legibility relies on the metaphor of territory-as-text, Lynchian legibility is perhaps best thought of as tracking by scent.

All three of the case studies take advantage of location sensing to generate data sets consisting of variables including timestamps, latitudes and longitudes, and derivatives such as speed, distance, or distribution. As a standardized, reductive, and essentially asemantic representation, these quantitative values also pinpoint a place, and *place* is layered with meaning. According to Waldo Tobler's first law of geography, "Everything is related to everything else, but near things are more related than distant things" (Tobler 1970). The geographic coordinate is an indexical point to a complex network of interrelations in which metric distance is not the only measure for expressing proximity. Regarded this way, spatial analysis becomes a qualitative endeavor.

### The Bottom-Up Perspective

In the Trash Track study, the observation of the waste system is limited to interpreting the recorded data points reported by the deployed location sensors. In the ideal scenario, a tracked item would report four to six locations per day. In reality it was typically less than that. The sparse data made it difficult to identify whether the report was sent from a facility or from the road. Localization artifacts made the reported location a matter of uncertainty. Because automatic algorithms for detecting stops at facilities, spatial clustering, and geocoding facilities require a certain amount of data to work reliably, analysis was mostly a manual process of collecting clues from various sources all the way from facility databases to waste management contracts. Demonstrating true dedication to waste forensics, my colleague David Lee spent part of his honeymoon on a road trip with his wife, exploring reported locations and visiting waste facilities in rural Oregon. But despite its sparsity, the data captured other aspects typically not included in official data sources, especially information about time and duration. Such information can be relevant for inferring carbon emissions of organic waste or strengthening evidence by matching the recorded trajectories with shipping documents. Overall, though, constructing evidence from sparse location data remains a precarious endeavor because multiple interpretations are possible.

The Forage Tracker study that took place in Brazil was initially motivated by a very Lynchian question: How do informal collectors read the urban environment, how do they find material, and which parameters inform their spatial decisions? Although the recorded GPS traces and observations in the cooperative recorded during the experiment were no more than anecdotal glimpses into a system that is opaque almost by definition, this time we had the opportunity to contextualize the traces with explanations from the collectors. Although each collector used and described different spatial strategies—focusing on specific materials and collecting from particular clients in particular areas—they all were influenced and constrained by the same parameters, including traffic, distance, and terrain, and most importantly the market prices of their goods.

In the third case study, citizen feedback systems offered a window into how residents perceived problems in their neighborhoods. The data was subjective, biased in many ways. The gravity of the described issue and the urgency expressed in the submitted report did not always correspond. Even if issues were perceived similarly, not every resident would decide to contact the city about it, and if they did, they used different media, ranging from letters to phone calls to smartphone reports. To some extent, citizens also read the city through the feedback app, perhaps alerted to issues in their neighborhood through the system. The city read the concerns of their constituents through the submissions captured by their constituent relationship management (CRM) system. In both cases, categories and interface design shaped how each party perceived the concerns. The study also demonstrated how an interface acts like a mirror: citizen reporters saw their role in the maintenance of infrastructure through their reports and the actions they triggered in a city department.

### The Top-Down Perspective

From the Scottian perspective of universal and reductive symbolic representations of the infrastructural landscape, the Trash Track study made it possible to integrate information across system boundaries such as state borders, service contract areas, transport modalities, or waste stream designations. Here, interpreting the recorded traces relied on the availability of official data sources.

In Forage Tracker, the difficulties in establishing legibility from above are manifest in the struggles by local, state, and national governments with collecting reliable data about the informal sector, which can serve as evidence to inform policy decisions. Unlike the abstract Scottian modernist state, the administrative levels within the informal sector are not the most powerful actors, and mandates requiring cooperatives and associations to report data to cities are executed to different extents and with different rigors, partially undermining the efforts of standardized data collection and benchmarking.

In the study of citizen feedback systems conducted in Boston, the effort to shape and establish legibility from above is manifest in the ongoing evolution of citizen feedback systems. The Scottian notion that administrative taxonomies influence reporting behavior becomes a bidirectional process of adaptation. The changing features and service types found within system interfaces bear witness to the iterative approach taken by local government to guide and shape their interactions with constituents and their attempts to reconcile internal structures and terminologies with the perception of issues by the citizens. Again, the boundary between constituents and officials is blurred by conscious design decisions in systems such as SeeClickFix that represent all parties in a similar way. This user parity is also seen in the fact that officials frequently use, out of convenience, the citizen feedback app to report issues.

### Data Formats and Visual Representations

Throughout this book, I have avoided a strict separation between data and visualization, as well as between sensing and displaying. The Trash Track data set was visualized in different ways, including animations, interactive graphics, static maps, and quick-and-dirty working models that used online mapping services. The multiple representations emerged from data explorations or were produced as public presentations. In Forage Tracker, data and maps were sometimes handwritten and sometimes created in a digital format, with one format often grafted onto another. Mapping a route involved recording a trace, printing it as a physical map, and annotating it manually during an interview. Occasionally the cooperatives produced maps of service areas and collection routes, but often the neighborhoods where they operated did not exist on official maps. In the case of citizen feedback apps, the maps used by the different systems were more consistent. All of them summarized reports as online markers. However, they all used different ways to represent users and to facilitate user interactions in the interface.

In the policy domain, visualization practice is often understood as the translation of predetermined messages into accessible visual forms. In research practices, transitions between data analysis and data visualization are more fluid, and the visualization practitioners who are often involved in data collection and analysis engage deeply with the characteristics and limitations of a particular data source. A failed visualization is often one that does not account for a data set's structures, error ranges, and biases. Data visualization artifacts are, like data sets, based on a codified symbolic language. In data analysis, visual and computational operations are often used and treated equivalently, and manipulating and transforming data sets usually involves exploring data as scatterplots that map the data into discrete or continuous color scales or arrange them in different spatial layouts. In all three case studies, visualization was an essential tool for the analysis and interpretation of the recorded spatial data.

### Issues of Data Analysis—the Stickiness of Context

At the beginning of this book, I introduced a definition of information as "data plus meaning," which implies that meaning is located external to the data artifacts. Critics of the Big Data paradigm point out that meaning is defined by the context of data collection—the precise conditions under which a data set was encoded—rather than through the data values (Drucker 2011). In fact, many geographic data sets collected in crowdsourced projects present themselves as an aggregation of decontextualized and underspecified location markers that were generated by an anonymous collective under unknown local conditions. The studies in this book demonstrate the difficulties of interpreting sensor data collected in unknown environments and deployed by participants with different motivations and interests. Without contextual information, we end up with close to nothing in our hands. Research that takes advantage of data generated by social media services such as Twitter struggles with similar issues because the demographics of the users who generated the data, along with their specific motivations and purposes for using the service, are often unknown. In such cases one could say that context is *sticky*; it cannot be ignored without diminishing the data value.

It turns out, however, that some data sources are less sticky than others, and they prove to be reliable proxies for modeling phenomena that are far removed from the original context of data collection. An example of such a data set is the workhorse of many geographers and economists, the data captured by the Operational Line Scanner (OLS) sensor on the satellites from the U.S. Defense Meteorological Satellite Program (DMSP). The geographic data grids, which show the nocturnal light emissions of cities and human activity, are used to model such diverse phenomena as economic

output, urbanization and poverty, resource footprints, and disease outbreaks (Sutton 1997; Henderson, Storeygard, and Weil 2009; Elvidge et al.
2011; Bharti et al. 2011). Interestingly, the data set is an entirely accidental byproduct of the military satellites built for measuring cloud cover for
reconnaissance missions (Hall 2001). Engineers discovered that their optical instruments were sensitive enough to register city lights, information
that closely correlates with energy utilization (Croft 1978; Welch 1980).
OLS data demonstrates the capacity of some data sources to transcend
original context if the apparatus of measurement and the context of observation are well understood and robust. Data generated through human
interactions, however, rarely fulfill this requirement. In both cases of sticky
and nonsticky, analysis requires a thorough attention to the context of
data generation.

### Data Interpretation beyond Truth and Bias
A second obstacle when reading socio-technical systems are the known and
hidden biases in data sets and research design constructs. The concept of
"data" has been conceptualized and scrutinized from different angles, but
the concept of "bias" is often taken for granted. "Bias" means a systematic
pattern of error or a deviation from the true mean of a distribution (Kitchin
2014, 14). In other words, the concept of bias implies a known truth. A
crowdsourced data set is biased in the sense that it originates from a self-
selected group of volunteers and is not a random sample drawn from the
larger population.

As the case study on citizen feedback systems in part III demonstrates,
there are many aspects of citizen-generated data sets that cannot be evaluated in terms of their accuracy. There is no canonical form of a citizen report
that could serve as a template to evaluate which themes are addressed, how
arguments are rhetorically framed, and how the received feedback influences reporting behavior.

Beyond how accurately citizen feedback represents a population and the
infrastructural issues they experience in their neighborhoods, the third case
study demonstrates that data sets can be used to investigate the dynamics
between citizens and cities by looking at how their interactions are influenced by the design factors of the mediating system. Because these data
sets never represent conditions that are stable in time, the design of the
reporting systems is constantly tweaked by cities, and users adjust their
behavior based on the feedback they receive. Computational social scientist David Lazer has shown that the predictive capability of the Google Flu

Trends service—which was designed to predict flu outbreaks based on user search terms—has degraded over time because users have started to change their search behavior in response to flu-related news they receive through the same search engine (Lazer et al. 2014). Such feedback phenomena are difficult to account for in terms of bias and truth without scrutinizing the dynamics of how the design of mediating technologies shapes the democratic discourse they facilitate.

## Legibility and the Transactionalization of Infrastructure

In the context of global waste systems, the need for infrastructure legibility is not difficult to demonstrate, considering the urgency of the environmental, public health, and equity issues as well as the lack of evidence necessary to make informed policy decisions. But if we expand this concept to other kinds of socio-technical systems, what is achieved by making infrastructure more legible? Am I overstating the importance of information and awareness? One might object that urban infrastructures are remarkably resilient even when they remain entirely illegible, due in part to the appropriations and improvisations of users.

As discussed throughout this book, infrastructure legibility has an important function for accountability, and I think there are reasons to assume that this dimension has become more important over the past decades. Contemporary urban systems are characterized by complex structures of governance and ownership. They are often run by a hybrid network of actors that include banks and pension funds, public institutions, private corporations, and community organizations.

As utility poles and street lights are retrofitted with networked sensors in many U.S. cities, the accountability dimension of infrastructure gets even more complicated. The party that owns and operates the sensors is not necessarily the same entity that owns the collected data and has accountability for what happens with the information. Questions of data life cycle, privacy, and public anonymity have to be solved in a complex network of accountability between all parties involved. At the same time, none of these aspects are legible to the pedestrian on the sidewalk, not even the fact that sensors are present and collecting data. Policies that regulate data sharing with third parties might change over the years, yet these changes do not have any consequences for how utility poles present themselves in the public space.

## Cryptocurrencies and the Transactionalization of the Public Good

As the Internet of Things (IoT) enters the public space, even minuscule infrastructure consumption becomes measurable, quantifiable, and ultimately billable. The capacity of IoT for micro-transactions between networked devices introduces another accountability-related aspect, which could be called the "transactionalization of infrastructure services." During the 2013 Turing Festival in Edinburgh, Mike Hearn, a former developer of the cryptocurrency Bitcoin, shared an idea for a future infrastructure he called TradeNet, which connects all existing objects and systems: "In this future scenario, the roads on which Jen is driving will have also become autonomous actors, doing trades with the car on TradeNet. They can submit bids to the car about how much they are going to charge to use them. If she is in a hurry, Jen can choose a road that is a bit more expensive but which will allow her to get into the city faster. Awesome, right?" (Hearn 2013).

It is not entirely clear what fuels Hearn's enthusiasm for this scenario, but let's assume it is the notion that the transportation infrastructure can be maintained entirely by billing users only for the "fair share" that corresponds to their service consumption. The governance of TradeNet is algorithmic, maintains a dynamic equilibrium, and adjusts prices based on market mechanisms to achieve an efficient system load. It is not difficult to compare this scenario to a "pay as you throw" model in the waste system, where each waste generator pays for the amount of waste generated. In both cases this model might have a positive environmental impact by affecting people's decisions to conserve resources.

But the TradeNet model also has interesting consequences for the status of infrastructure as a public good. As discussed earlier in this book, when service consumption can be accurately measured and billed with few transaction costs, exclusion becomes more feasible and the common good becomes a private good. In the case of Bitcoin, this is not without irony since the blockchain, the underlying infrastructure necessary to verify Bitcoin micro-transactions, is a common good itself. The system would not function without the contributions of Bitcoin enthusiasts who run the transaction-verifying nodes by contributing their own time, hardware, and electricity.

Bitcoin is a paradoxical commons. From the outside, its network is presented as a neutral and incorruptible self-governing algorithmic system. Bitcoin "is regulated, only by mathematics instead of politicians," according to a common argument by supporters (Voorhees 2012). In this perspective, governance is seen as a form of housekeeping, required only to enforce existing contracts. "Just as robots have helped the world reduce menial physical labor, so cryptocurrency technology now gives us the tools to automate the menial labor of bureaucracy. Optimistically, the entirety of humanity will benefit as a result" (Barski and Wilmer 2014).

Not only is this concept of governance very different from the one used in this book, it is also inconsistent with what happens in the Bitcoin community itself. For more than two years, as the digital currency has gained popularity, the community has been deeply divided by the "block size debate," a controversy around the appropriate bandwidth of the blockchain determined by the atomistic size of its blocks storing Bitcoin transactions. This seemingly trivial technical detail has wide-ranging, even geopolitical implications for the distribution of power among the participating actors since most Bitcoin mining activities are concentrated in China.

From the perspective of infrastructure legibility, Bitcoin is black-boxed. It presents itself to the outside as a transparent, transactional, and incorruptible algorithmic system of governance. But in fact, it is dominated by the same kinds of controversies and politics that shape most other sociotechnical systems.

### The Aesthetics of Transactionalization

The schizophrenic aesthetics of systems that represent themselves outwardly as conceptually simple and algorithmically precise, hiding the messy negotiations necessary to keep the system running, are not limited to the Bitcoin network. From platform companies such as Uber to search engines, most digital services employ similar design choices. The minimalistic interface of Google's search engine hides the company's constant tinkering with its search algorithm to neutralize attempts by outsiders to manipulate the search results. The Uber smartphone app looks the same in every city of the world, hiding the fact that the company often has to negotiate with each city government to comply with local regulations. Contributors to crowdsourcing platforms are invisible and abstracted into a unified digital service API as ironically acknowledged by Amazon's Mechanical Turk platform, which in name refers to an eighteenth-century faux automaton: a human chess player pretending to be a machine.

### Algorithm Awareness

All of these phenomena introduce new challenges for reading infrastructures. In the case of Bitcoin, the controversies and discussions around the protocol take place in the public, the code is open source, and all changes are extensively discussed in the community. Kevin Hamilton and colleagues have investigated issues of "algorithm awareness," meaning the degree to which everyday users are aware of the invisible algorithms that determine their online experience. As the authors explain, "Algorithms are buried not only outside of human perception, but behind walls of intellectual property" (Hamilton et al. 2014, 632). Questions of algorithm awareness gained public attention when Facebook researchers manipulated newsfeeds without user consent to study how emotions spread through the network (Kramer 2012), raising the concern over how such practices could be deployed in the context of national elections. Another example is the "right to be forgotten" legislation of the European Union (Mantelero 2013), which allows individuals to request the exclusion of their person from Internet search results to protect them from abuse. At the same time, this right raises concerns about the possibility of manipulating the online representation of public figures.

Algorithmic governance brings to the fore certain dilemmas of *algorithmic accountability* (Diakopoulos 2014). Some forms of algorithmic governance work only as long as they remain secret. Early search engines such as AltaVista failed because users reverse-engineered the search algorithm, making their own sites more visible while diminishing the quality of results for everyone else. For similar reasons, algorithms used to calculate credit scores remain secret, giving only vague indications about which factors are considered. In urban space, algorithmic modes of governance produce what Steve Graham describes as "software-sorted geographies" that can manage the visibility of points of interest in online maps, direct users by way of navigation systems, or spatially adjust service rates (Graham 2005).

In the space of algorithmic governance, infrastructure legibility is first of all an issue of accountability, a question of integrating the nature and function of algorithms into the democratic discourse. Making the function of algorithms legible raises several dilemmas for which full transparency is not always a solution. In the last section of this book, I outline design principles that allow us to navigate the dilemmas and complexities involved in cyber-physical infrastructures.

## A Case for Accountability-Oriented Design

Throughout this book, I have argued that design and governance are closely related. First, design involves many aspects of governance. From architecture to smartphone user interfaces, design regulates behavior and frames issues in certain ways. Second, governance also shares similarities with design. Setting policies and negotiating rules requires reconciling contradictory factors and making small adjustments over multiple iterations.

In the preceding chapters, I was concerned with different aspects and practices of dissecting and reverse-engineering waste systems. I will conclude by proposing provisional design principles that adapt the preceding discussions for contemporary cyber-physical urban infrastructures. My principles of accountability-oriented design run counter in many ways to the traditional ideas about "good design." The functionalist design principles of clarity and simplicity that define current information design practices aim at reducing complexity to its essence (Rams 1984). But in the case of the messy, ambiguous, and sometimes paradoxical reality of infrastructural systems, such essentialism is futile. As Don Norman and Pieter Jan Stappers argue, the human mind is not well equipped to investigate socio-technical systems since we tend to look for simple, reductive models, a tendency reinforced by minimalist design heuristics (Norman and Stappers 2015).

To acknowledge the nature of complex socio-technical systems, a different set of design heuristics is needed. Such an approach would avoid deceptively simple and reductive representations, calling attention instead to the multiple perspectives on infrastructures. What could be called "accountability-oriented design" calls attention to issues of governance and the role of design as an agent that regulates human behavior and system interactions. In the remainder of this chapter, I discuss a proposal for accountability-oriented design that:

1. calls attention to issues of governance;
2. considers the systems around objects;
3. spans physical and informational domains;
4. enables public discourse by managing visibility;
5. is seamful rather than seamless;
6. does not explain; it shows;
7. is rich and redundant; and
8. acknowledges its own limitations.

**Accountability-Oriented Design Calls Attention to Issues of Governance.**
Accountability-oriented design is contextual. It considers a specific situation and a particular group of constituents to provide governance-related information where and when it is needed. In the earlier example of the sensor-equipped utility pole, an accountability-oriented design approach could mean alerting pedestrians to the presence of the sensor and pointing

to online resources that provide details about the sensor's data governance, such as how long information is stored and who has access to it.

An example of accountability-oriented design is what political scientist Dieter Zinnbauer terms "ambient accountability," defined as "all efforts that seek to shape, use and engage systematically with the built environment and public places and the ways people experience and interact in them, in order to further transparency, accountability and integrity of public authorities and services" (Zinnbauer 2012). As an example of ambient accountability, Zinnbauer cites a construction site display, such as those legally required in many countries. The display identifies the architect, the client, and the construction company, the beginning and anticipated end dates, and the budget and funding sources if the project is public. If the construction site appears abandoned or presents safety hazards, it is important to have this information presented on site rather than hidden in institutional databases. Maintenance logs on machines and cleaning schedules in public bathrooms fall into the same category.

Ambient accountability also assists in educating the public. "Know your rights" murals can be found in parts of New York City, especially in areas with large ethnic minority and African-American populations. These murals inform people about their rights relevant to encounters with police—for example, that it is legal and encouraged (by the mural at least) to film officers on duty during arrests. Ambient accountability also includes actions of public shaming, such as the inflatable rats erected by members of U.S. labor unions in front of businesses that do not use unionized labor. But ambient accountability can also be more subtle in its expressions, for example in the way public officials represent themselves in their own offices.

### Accountability-Oriented Design Considers the Systems around Objects.
Just like the utility pole, most objects are embedded in larger systems. An accountability-oriented design approach looks beyond the boundaries of the object and considers how its different roles in the surrounding systems can be communicated through design. This can involve simple gestures, such as the practice of labeling waste bins "landfill" to designate the larger system the bin is part of. Mandatory product-labeling requirements, such as the disclosure of health effects of food or cigarettes, the environmental impact of packaging, and the exact meaning of terms such as "compostable" and "recyclable" are subjects of ongoing battles between regulators and industry precisely because they call attention to controversies in the larger systems of food production and manufacturing.

**Figure C.1**
Know Your Rights mural in Bushwick, Brooklyn by artist Dasic Fernández. Screen-shot from Google Street View, reproduced under Google Maps fair-use policy.

Accountability-oriented design is relational; it concerns the ways in which objects announce themselves to their surroundings. The shutter sound of cell phone cameras is not just a nostalgic reference to analog cameras, but indicates to the people in proximity that a photo has been taken. For this reason, some countries require phone manufacturers to include this sonic signifier, acknowledging that taking a photo is inappropriate in some situations.

### Accountability-Oriented Design Spans Physical and Informational Domains.

What can be communicated by attaching a physical label and defining a designated icon is limited. Accountability-oriented design can address this by making sure that sensors in public spaces are both physically and virtually identifiable. If through acceptable practice a waste bin collects data about pedestrian activity by skimming hardware addresses from personal devices, it should be possible for people to connect to this sensor through their own phones to access accountability-related information, such as the email addresses of those responsible for safeguarding the collected data.

Both physical and digital components are necessary because they are contingent upon each other. The open data movement frequently invokes the notion of "digital public space," a web server where public data sets can be accessed. But despite this rhetoric, digital and physical spaces are not equivalent in their "publicness." According to urban writers Jane Jacobs and Richard Sennett, the involuntary exposure to diversity is a central aspect of the public space where one cannot choose who one runs into (Jacobs 1961; Sennett 1970). This is not the case in digital space where chance encounters are less likely and often managed by a filter bubble. The only people who will be able to utilize open data sets are those who actively seek them, know where to look for them, and know how to work with them. The idea of ambient accountability could be instrumental for connecting open data resources to the physical places where they are relevant. QR-codes, the two-dimensional barcodes that can be read by a smartphone, have become ubiquitous in public space, but they are not human readable, so they can be augmented with relevant, human-readable information.

### Accountability-Oriented Design Enables Public Discourse by Managing Visibility.

In part III of this book, I discussed how the perception of the user in citizen feedback systems is managed by the configurations of the interface. This is not necessarily a deceptive practice, but a decision every designer of social interfaces has to make. Every design decision has consequences for user behavior, including many that cannot be anticipated. Accountability-oriented design suggests that a designer should be aware of his or her own role in regulating and shaping behavior.

But managing visibility also means considering the back-ends of protocols and software licenses that regulate the visibility of technical arrangements. The synergy and kinship between open standards, open source software, and participatory democracy are recurrent themes in many public sector projects. Since 2003, the city of Munich has migrated over fifteen thousand computers to the open-source operating system Linux. Other examples of open source in government include the adoption of the OpenDocument standard for all Massachusetts state entities in 2005, the partially crowdsourced Icelandic constitution reform, and the citizen-written transparency law in Hamburg, Germany (Shah, Kesan, and Kennis 2008; Landemore 2015; Verein f. mehr Demokratie 2012).

### Accountability-Oriented Design Is Seamful Rather than Seamless.

The notion of seamful design has been discussed in several instances in this book. At this point, I want to clarify that *seamful design* does not necessarily mean making an experience deliberately inconvenient.

Private ride-sharing services such as Uber and Lyft are increasingly integrated into public transportation systems. Navigation apps that combine real-time data from multiple modes of transportation make trips that involve public and private modes of transportation an almost seamless experience. At the same time, the boundaries between public and private services can become obscured. As local governments and transit authorities seek collaborations with ride-sharing companies, complex questions of data ownership and accountability arise. Seamful design in this context could mean disentangling the accountability relationships in public-private partnerships.

### Accountability-Oriented Design Does Not Explain; It Shows.

Connecting to a public Wi-Fi hotspot typically involves scrolling through several pages of a usage agreement that details the legal implications of connecting to the Internet through the provided connection. However obscure and misleading their language may be, these agreements represent an important accountability mechanism. Nevertheless, few people read them in places like airports where perusing detailed explanations of a socio-technical system is out of the question. In addition to the agreements, logging onto the hotspot could provide an abstracted, socially translucent representation of the activity of other users who are currently connected to the same system.

Reading clues and traces requires less effort than reading a text or decoding a symbolic language. This is the rationale behind *ambient displays* residing at the periphery of attention (Wisneski et al. 1998; Offenhuber 2008). The LEDs on an Internet router are symbolic representations of technical states that are unknown and meaningless to most users. They do not explain the function of the device or, more generally, the TCP/IP network. They nevertheless convey a sense of activity that most people are able to understand. If the LEDs remain relatively calm for hours, only to burst into frantic activity in the middle of the night, one might get curious about whether someone is trying to break into the network or the computer is simply performing a regular update.

**Accountability-Oriented Design Is Rich and Redundant.**
In part III of this book, I contrasted the rich and messy appearance of "dirty visualizations" created by collaborators using whatever software and tools were at hand, with the polished, minimalist designs of professional information designers. Accountability-oriented design communicates on multiple levels, addresses different contexts and situations, and therefore necessarily involves redundancies.

**Accountability-Oriented Design Acknowledges Its Own Limitations.**
Making things legible by sharing and publishing information is no universal remedy. Accountability-oriented design also requires a realistic and critical reflection on what can and should be addressed through design and data collection. Often, vulnerable groups have to withhold information to avoid having more powerful groups take advantage of them. Many things have to remain hidden. A whistleblower platform depends on the trust that the identity of the whistleblower remains anonymous. Transparency platforms call attention to failures rather than successes, which can be instrumentalized by political opponents. At the same time, radical transparency is an effective way of obfuscating relevant signals within a torrent of noise.

In many cases, the responsible solution is not to collect data at all. An accountability-oriented design approach is therefore not only concerned with showing what happens, but also with what does not happen, as secrets that should not be disclosed should be safeguarded. As the case study of Brazilian recycling cooperatives described in part II has shown, attempts at making informal practices legible through technology sometimes arrive at the conclusion that these practices should remain illegible.

## Conclusion

In this book, I have described infrastructure legibility as structure and process, presence and social practice, governance and the civic self. I have used legibility in the context of data collection, experience and awareness, visual communication, and narratives for guidance. Legibility involves reading infrastructure through material and digital interfaces as well as through human practices and performances in physical and informational spaces. The aspects of infrastructure legibility that I have described are not conclusive categories; they are heuristics that make no claim to completeness or universality.

In an environment of increasingly mediated infrastructure, urban planners can learn from design disciplines that deal from the outset with the experience of infrastructure and its implications. The confusing and contradictory recycling ordinances and bins with a variety of shapes and colors are two obvious indications that the interfaces of waste systems suffer from a lack of attention to design. But designers are also frequently oblivious to the political nature of their artifacts and can benefit from lessons offered by social sciences and the humanities.

Waste systems are fitting exemplars of heterogeneous, illegible, and contested infrastructures. The archaic, physical nature of waste systems forces us to think closely about the process of observing a particular aspect, encoding it into data sets, and constructing evidence to inform policy decisions or enforce environmental laws. The implications of this study are not limited to waste infrastructures. To prevent the contents of open data portals and transparency initiatives from becoming informational waste that clutters the arena of public discourse, we must pay attention to the experience of information, which is not simply there and ready-to-hand, but something that needs to be made.

# Notes

## Introduction

1. The accusative supine of *dare*, to give.

2. From the lower boundary of space at an altitude of 100 kilometers the former landfill can certainly be recognized, but so can many other human-made structures.

## Prologue to Part I

1. For more examples, see http://senseable.mit.edu/.

## Chapter 1

1. For Comprehensive Environmental Response, Compensation, and Liability Act Information System, discussed later in this chapter.

2. Access to various databases is available through EPA's Environfacts portal, https://www3.epa.gov/enviro.

3. EPCRA technically is an amendment to CERCLA, but is usually referred to by its own name.

4. Criminal law, applicable for environmental crimes, requires establishing conclusive evidence, while environmental law, for injunctions and corrective action, only requires demonstrating a significant probability.

5. From a personal conversation with the author 2015.

6. For an example of such crowdsourced applications, see http://www.inaturalist.org/.

7. From an interview conducted by the author with Shannon Dosemagen, executive director of the Public Lab, May 29, 2013.

8. For an example of similar services, see http://bigbelly.com/.

9. A circumstance that information activists such as Pablo Rey Mazon try to address by mapping and adding landfills and waste facilities to OpenStreetMap, http://publiclab.org/profile/pablo.

10. See http://maps.nyc.gov/snow.

## Chapter 2

1. Some aspects of the analysis, arguments, and conclusion discussed in this chapter were published in Offenhuber et al. 2012.

2. See http://en.beidou.gov.cn/.

3. See https://www.glonass-iac.ru/en.

4. See http://www.esa.int/Our_Activities/Navigation.

5. See https://www.thetileapp.com/.

6. Bootstrapping refers to a statistical method for acquiring confidence intervals based on small samples.

7. The spot market is a financial exchange where commodities are traded immediately or "on the spot."

8. From an email conversation in July 2010. The text and context have been anonymized.

9. The name refers to the global standards organization ASTM International, initially an acronym for American Society for Testing and Materials. See https://www.astm.org/.

10. Converted from BTU/short tons used in the WARM documentation.

11. The carbon dioxide equivalent is a common term to describe the global warming impact of different greenhouse gases.

12. Converted from the transport emission factor of 0.04kg $CO_2$e/ton-mile, based on a fuel efficiency value of 0.0118 gal/ton-mile or 0.027 liters/ton-km (Scharfenberg, Pederson, and Choate 2004, 2).

## Epilogue to Part I

1. The acronym originally stood for "Commercial Hazardous Waste Management Evaluation Group," but it is no longer explained in any publications.

## Chapter 3

1. See http://marcolombia.co/. CEMPRE is originally a Brazilian initiative that has expanded to other Latin American countries.

2. Gallianzo Avisa, a project by the Peruvian government tracking vultures to identify illegal dumpsites; see http://www.gallinazoavisa.pe/.

3. A research project for visual litter recognition. See http://en.research.pirika.org/.

4. See for example the U.S. initiative Local Data, http://localdata.com/, or Make My Island, a UNDP initiative for mapping dumpsites in the Maldives, http://www.makemyisland.mv/.

5. See http://wecyclers.com/.

6. See http://www.kabadiwallaconnect.in/.

7. See http://www.pedalpeople.coop/.

**Chapter 4**

1. In Portuguese, Política Nacional de Resíduos Sólidos (PNRS), law no. 12,305/2010.

2. In Portuguese, *Movimento Nacional dos Catadores de Materiais Recicláveis* (MNCR).

3. In Portuguese, Compromisso Empresarial para Reciclagem, or Business Commitment to Recycling, http://www.cempre.org.br/.

4. See http://www.cataacao.org.br/.

5. See http://www.avina.net/.

6. See http://www.iadb.org/.

7. See http://www.igainstitute.com/.

8. Some data and findings discussed in this chapter were previously published in Offenhuber and Lee 2012. The Forage Tracking project website is http://senseable.mit.edu/foragetracking.

9. In Portuguese, *Pontos de Entrega Voluntária* (PEV).

**Addendum**

1. See http://www.rapidresults.org/map-location/brazil-informal-waste-sector and http://web.archive.org/web/20131109145014/http://www.cataacao.org.br/institucional/programa.

2. In English: Project on the Environment and Citizenship.

**Prologue to Part III**

1. See http://de.guttenplag.wikia.com/wiki/Datei:Thumb_xxl.png.

2. See http://driven-by-data.net/2011/03/01/plagiarism.html.

3. See https://petabencana.id/map/jakarta.

## Chapter 5

1. See, for example, https://www.wunderground.com/weatherstation/overview.asp.

2. A Linux distribution is an operating system consisting of various software packages that is based on the Linux kernel.

3. See the archived page: http://web.archive.org/web/20050730015325/http://www .advocatesforrasiej.com/wefixnyc.

4. From a seminar at Northeastern University with Nigel Jacob and Chris Osgood from the city of Boston, April 2015.

5. Ibid.

6. Tellingly, the director initially wanted to call his film *Tarzan vs. IBM*.

## Chapter 6

1. In 2015, the service was renamed Boston 311; http://www.cityofboston.gov/311.

2. Some data and findings discussed in this chapter have been presented at the 2014 Workshop on Big Data and Urban Informatics (Offenhuber 2014).

3. Parts of the study discussed in this chapter have been published in Offenhuber 2015.

4. See http://www.open311.org/.

5. From an unpublished interview by the author, 2011.

6. From an unpublished interview by the author with an analyst for the city of New York, 2012.

7. From an unpublished interview by the author, 2014.

8. From an unpublished interview by the author, 2012.

9. See http://snowstats.boston.gov/.

10. As of September 2016.

11. As of September 2016, accessible on https://311.boston.gov/.

12. From an unpublished interview by the author, 2014.

# References

Agência Estadual de Meio Ambiente. 2012. "Plano Estadual de Resíduos Sólidos." Recife, Pernambuco. http://www.cprh.pe.gov.br/downloads/PlanoResiduoSolido _FINAL_002.pdf.

Alcoa. 2010. "Alcoa Launches Aluminum Can Recycling App for iPhone." *Alcoa News Releases*. October. http://web.archive.org/web/20101119174409/http://alcoa .com/global/en/news/news_detail.asp?pageID=20101026006720en&newsY ear=2010.

Anonymous, and Martin Kotynek. 2013. "Reflections on a Swarm." In *Accountability Technologies—Tools for Asking Hard Questions*, ed. Dietmar Offenhuber and Katja Schechtner, 76–83. Vienna: Ambra V.

Arnstein, Sherry R. 1969. "A Ladder of Citizen Participation." *Journal of the American Institute of Planners* 35 (4): 216–224.

Ayoub e Silva, Ana Carolina Corberi Famá, Manuela Prado Leitão, and Patrícia Faga Iglecias Lemos. 2014. "Packaging and Information: The Importance of Environmental Information in the Environmental Challenges of Sustainable Waste Management." In *Design Waste & Dignity*, ed. Maria Cecília Loschiavo dos Santos, Stuart Walker, Sylmara Lopes, and Francelino Gonçalves Dias, 187–203. São Paulo, SP: Editora Olhares.

Baker, M. N. 1901. "Condition of Garbage Disposal in United States." *Municipal Journal and Engineer* 11:147–148.

Barski, Conrad, and Chris Wilmer. 2014. *Bitcoin for the Befuddled*. San Francisco: No Starch Press.

Bartels, Koen P. R., Guido Cozzi, and Noemi Mantovan. 2013. "'The Big Society,' Public Expenditure, and Volunteering." *Public Administration Review*, 73 (2): 340–351. doi:10.1111/puar.12012.

Basel Action Network. 2016. "E-Trash Transparency Project." *Basel Action Network*. http://www.ban.org/trash-transparency/.

Bateson, Gregory. 1972. *Steps to an Ecology of Mind: Collected Essays in Anthropology, Psychiatry, Evolution, and Epistemology*. Chicago: University of Chicago Press.

Bazo Soluções. 2015. "CataFácil." http://www.catafacil.com.br/index.html.

Ben-Shahar, Omri, and Carl E. Schneider. 2014. *More Than You Wanted to Know: The Failure of Mandated Disclosure*. Princeton, NJ: Princeton University Press.

Berkowitz, Ben. 2009. "SeeClickFix Releases First iPhone App." *WeMedia.com*. August 14. http://wemedia.com/2009/08/14/seeclickfix-releases-first-iphone-app/.

Bharti, N., A. J. Tatem, M. J. Ferrari, R. F. Grais, A. Djibo, and B. T. Grenfell. 2011. "Explaining Seasonal Fluctuations of Measles in Niger Using Nighttime Lights Imagery." *Science* 334 (6061): 1424–1427. doi:10.1126/science.1210554.

Bijker, Wiebe E., Thomas P. Hughes, Trevor Pinch, and Deborah G. Douglas. 1987. *The Social Construction of Technological Systems: New Directions in the Sociology and History of Technology*. Cambridge, MA: MIT Press.

Birkbeck, Chris. 1979. "Garbage, Industry, and the 'Vultures' of Cali, Columbia." In *Casual Work and Poverty in Third World Cities*, ed. Ray Bromley and Chris Gerry. Chichester: John Wiley & Sons.

Blei, David M. 2012. "Probabilistic Topic Models." *Communications of the ACM* 55 (4): 77–84.

Bleiwas, Donald I., and Thomas Dudley Kelly. 2001. "Obsolete Computers: 'Gold Mine,' Or High-Tech Trash?: Resource Recovery from Recycling." U.S. Department of the Interior, U.S. Geological Survey. http://pubs.usgs.gov/fs/fs060-01/.

Bonner, Sean. 2012. "Safecast." In *Inscribing a Square: Urban Data as Public Space*, ed. Dietmar Offenhuber and Katja Schechtner, 50–54. Vienna; New York: Springer.

Borden, Ed, and Adam Greenfield. 2011. "YOU Are the 'Smart City.'" Archived. *Pachube Blog*. June 30. http://web.archive.org/web/20110701132439/http://blog.pachube.com/2011/06/you-are-smart-city.html.

Boustani, Avid, Lewis Girod, Dietmar Offenhuber, Rex Britter, Malima I. Wolf, David Lee, Stephen B. Miles, Assaf Biderman, and Carlo Ratti. 2011. 'Investigation of the Waste-Removal Chain through Pervasive Computing." *IBM Journal of Research and Development* 55 (1/2): 11:1–11:11.

Bowker, Geoffrey C. 1994. *Science on the Run: Information Management and Industrial Geophysics at Schlumberger, 1920–1940*. Cambridge, MA: MIT Press.

Bowker, Geoffrey C., and Susan Leigh Star. 1999. *Sorting Things Out*. Cambridge, MA: MIT Press.

Boyd, Danah, and Kate Crawford. 2012. "Critical Questions for Big Data: Provocations for a Cultural, Technological, and Scholarly Phenomenon." *Information Communication and Society* 15 (5): 662–679. doi:10.1080/1369118X.2012.678878.

Boyle, Richard. 1995. *Towards a New Public Service*. Dublin: Institute of Public Administration.

Brabham, Daren C. 2012. "The Myth of Amateur Crowds." *Information Communication and Society* 15 (3): 394–410. doi:10.1080/1369118X.2011.641991.

Brandeis, Louis Dembitz. 1914. *Other People's Money: And How the Bankers Use It*. New York: F.A. Stokes.

Brazil. 2006. "Simples Nacional." http://www8.receita.fazenda.gov.br/SIMPLES NACIONAL/Default.aspx.

Brazil. 2010. "Lei 12305/2010—Residuos Solidos." http://www.trabalhoseguro.com/ Leis/Lei_12305_2010.html.

Brinton, Willard C. 1914. *Graphic Methods for Presenting Facts*. New York: The Engineering magazine company.

Brown, Phil. 1987. "Popular Epidemiology: Community Response to Toxic Waste-Induced Disease in Woburn, Massachusetts." *Science, Technology & Human Values* 12 (3/4): 78–85.

Brown, Phil. 1997. "Popular Epidemiology Revisited." *Current Sociology* 45 (3): 137–156. doi:10.1177/001139297045003008.

Buchanan, Richard. 1985. "Declaration by Design: Rhetoric, Argument, and Demonstration in Design Practice." *Design Issues* 2 (1): 4–22. doi:10.2307/1511524.

Budlender, Debbie. 2011. "Statistics on Informal Employment in Brazil." WIEGO Statistical Brief No. 2. WIEGO (Women in Informal Employment: Globalizing and Organizing). http://wiego.org/publications/statistics-informal-employment-brazil.

Buell, Ryan W., Ethan Porter, and Michael I. Norton. 2014. "Surfacing the Submerged State: Operational Transparency Increases Trust in and Engagement with Government." SSRN Scholarly Paper ID 2349801. Rochester, NY: Social Science Research Network. https://papers.ssrn.com/abstract=2349801.

Bullard, Robert D. 2000. *Dumping in Dixie: Race, Class, and Environmental Quality*. 3rd ed. Boulder, CO: Westview Press.

California, Department of General Services. 2000. "Non-Emergency Number Pilot Programs." Sacramento, CA: Department of General Services. http://www .911dispatch.com/reference/cal311report.pdf.

Callon, M., and B. Latour. 1981. "Unscrewing the Big Leviathan: How Actors Macro-Structure Reality and How Sociologists Help Them to Do So." *Advances in Social Theory and Methodology: Toward an Integration of Micro-and Macro-Sociologies*, ed. K. Knorr-Cetina and A. V. Cicourel, 277–303. London: Routledge.

Cardwell, Diane. 2002. "Bloomberg Plans Quick Start of Citywide 311 Phone System." *New York Times*, February 1. http://www.nytimes.com/2002/02/01/

nyregion/bloomberg-plans-quick-start-of-citywide-311-phone-system.html
?pagewanted=all&src=pm.

Casson, Tony, and Patrick S. Ryan. 2006. "Open Standards, Open Source Adoption
in the Public Sector, and Their Relationship to Microsoft's Market Dominance."
SSRN Scholarly Paper ID 1656616. Social Science Research Network. http://
papers.ssrn.com/abstract=1656616.

Castells, Manuel, and Alejandro Portes. 1989. "World Underneath: The Origins,
Dynamics, and Effects of the Informal Economy." In *The Informal Economy: Studies in
Advanced and Less Developed Countries*, ed. Alejandro Portes, Manuel Castells, and
Lauren A Benton, 11–37. Baltimore, MD: Johns Hopkins University Press.

Cavill, Sue, and Muhammad Sohail. 2004. "Strengthening Accountability for
Urban Services." *Environment and Urbanization* 16 (1): 155–170. doi:10.1177/
095624780401600113.

CBS News. 2008. "Following the Trail of Toxic E-Waste." *CBS News*. http://www
.cbsnews.com/news/following-the-trail-of-toxic-e-waste.

Chalmers, Matthew, and Areti Galani. 2004. "Seamful Interweaving: Heterogeneity
in the Theory and Design of Interactive Systems." In *Proceedings of the 5th Conference
on Designing Interactive Systems: Processes, Practices, Methods, and Techniques*, 243–252.
New York: ACM. http://dl.acm.org/citation.cfm?id=1013115.

Charnov, E. L. 1976. "Optimal Foraging, the Marginal Value Theorem." *Theoretical
Population Biology* 9 (2): 129–136.

Chen, Martha. 2012. "The Informal Economy: Definitions, Theories and Policies." 1.
WIEGO Working Paper. WIEGO (Women in Informal Employment: Globalizing and
Organizing), Manchester.

Chen, Martha A. 2016. "Technology, Informal Workers and Cities: Insights from
Ahmedabad (India), Durban (South Africa) and Lima (Peru)." *Environment and
Urbanization* 28 (2): 405–422. doi:10.1177/0956247816655986.

Chow, William, David Block-Schachter, and Samuel Hickey. 2014. "Impacts of
Real-Time Passenger Information Signs in Rail Stations at the Massachusetts Bay
Transportation Authority." *Transportation Research Record: Journal of the Transporta-
tion Research Board* 2419 (December): 1–10. doi:10.3141/2419-01.

City of Chicago. 2013. "Chicago 311 History." http://www.cityofchicago.org/city/
en/depts/311/supp_info/311hist.html.

Clapp, Jennifer. 2002. "The Distancing of Waste: Overconsumption in a Global
Economy." In *Confronting Consumption*, ed. Thomas Princen, Michael F. Maniates,
and Ken Conca, 155–176. Cambridge, MA: MIT Press.

Code for America. 2016. "Who We Are." *Code for America.* http://www .codeforamerica.org.

Colab. 2010. "The Green Grease Project." *MIT Community Innovators Lab.* http:// greengrease.scripts.mit.edu/home/.

Coleman, D. J., Y. Georgiadou, and J. Labonte. 2009. "Volunteered Geographic Information: The Nature and Motivation of Produsers." *International Journal of Spatial Data Infrastructures Research* 4:332–358.

Comber, Rob, Anja Thieme, Ashur Rafiev, Nick Taylor, Nicole Krämer, and Patrick Olivier. 2013. "BinCam: Designing for Engagement with Facebook for Behavior Change." In *Human-Computer Interaction—INTERACT 2013,* ed. Paula Kotzé, Gary Marsden, Gitte Lindgaard, Janet Wesson, and Marco Winckler, 99–115. Lecture Notes in Computer Science. Berlin: Springer. doi:10.1007/978-3-642-40480-1_7.

Commonwealth of Massachusetts. 2012. "Citizens Commonwealth Connect App." *Administration and Finance.* http://www.mass.gov/anf/budget-taxes-and -procurement/community-innovation-challenge-grant/citizens-commonwealth -connect-app.html.

Connett, Paul, and Helen Spiegelman. 2013. "Multimaterial Curbside Recycling and Producer Responsibility." In *The Zero Waste Solution: Untrashing the Planet One Community at a Time,* 245–252. White River Junction, VT: Chelsea Green Publishing.

Consonni, Stella. 2013. "Brazil's Incoming E-Waste Recycling Regulations Explained." *Waste Management World,* March 18. https://waste-management-world .com/a/brazils-incoming-e-waste-recycling-regulations-explained.

Cooke, Bill, and Uma Kothari. 2001. *Participation: The New Tyranny?* London, New York: Zed Books.

Croft, Thomas A. 1978. "Nighttime Images of the Earth from Space." *Scientific American,* July. http://www.scientificamerican.com/article/nighttime-images-of-the-earth -from/.

Davis, Stacy Cagle, Susan W. Diegel, and Robert Gary Boundy. 2009. "Transportation Energy Data Book: Edition 28." Technical Report ORNL/TM-2009/149. Oak Ridge National Laboratory (ORNL).

Deerwester, Scott C., Susan T. Dumais, Thomas K. Landauer, George W. Furnas, and Richard A. Harshman. 1990. "Indexing by Latent Semantic Analysis." *Journal of the Association for Information Science and Technology* 41 (6): 391–407.

De Jong, Frieda, and Karen Mulder. 2012. "Citizen-Driven Collection of Waste Paper (1945–2010): A Government-Sustained Inverse Infrastructure." In *Inverse Infrastructures: Disrupting Networks from Below,* ed. Tineke M. Egyedi and Donna C. Mehos, 83–102. Cheltenham, UK; Northhampton, MA: Edward Elgar Publishing.

Deloitte. 2015. "Supply Chain Forensics | Deloitte US | Forensic Investigations." *Deloitte United States.* https://www2.deloitte.com/us/en/pages/advisory/solutions/ supply-chain-forensic-services.html.

Deming, W. Edwards. 1993. *The New Economics for Industry, Government, Education.* Cambridge, MA: Massachusetts Institute of Technology, Center for Advanced Engineering Study.

de Soto, Hernando. 1989. *The Other Path: The Invisible Revolution in the Third World.* New York: Basic Books.

Desouza, Kevin C., and Akshay Bhagwatwar. 2012. "Citizen Apps to Solve Complex Urban Problems." *Journal of Urban Technology* 19 (3): 107–36. doi:10.1080/10630732. 2012.673056.

Deterding, Sebastian, Miguel Sicart, Lennart Nacke, Kenton O'Hara, and Dan Dixon. 2011. "Gamification. Using Game-Design Elements in Non-Gaming Contexts." In *PART 2—Proceedings of the 2011 Annual Conference Extended Abstracts on Human Factors in Computing Systems*, 2425–2428. CHI EA '11. New York: ACM. doi:10.1145/ 1979482.1979575.

Diakopoulos, Nick. 2014. "Algorithmic Accountability Reporting: On the Investigation of Black Boxes." Tow/Knight Brief. Tow Center for Digital Journalism, Columbia University, New York.

Diário de Pernambuco. 2013. "Pró-Recife Realiza Coleta Diariamente." May 5. http:// web.archive.org/web/20131222221020/http://www.folhape.com.br/cms/opencms/ folhape/pt/edicaoimpressa/arquivos/2013/05/05_05_2013/0017.html.

Dias, Sonia. 2009. "Overview of Legal Framework for Social Inclusion in Solid Waste Management in Brazil." http://wiego.org/sites/wiego.org/files/publications/files/ Dias_Brazil_Legal_framework_social_inclusion_waste_0.pdf.

Dirks, Susanne, and Mary Keeling. 2009. "A Vision of Smarter Cities: How Cities Can Lead the Way into a Prosperous and Sustainable Future." IBM Global Services Executive Report, Somers, NY.

Dodge, Martin, and Rob Kitchin. 2004. "Flying through Code/space: The Real Virtuality of Air Travel." *Environment & Planning A* 36 (2): 195–212.

Donath, Judith. 2014. *The Social Machine: Designs for Living Online.* Cambridge, MA: MIT Press.

Dosemagen, Shannon, Matthew Lippincott, Liz Barry, Don Blair, and Jessica Breen. 2013. "Civic, Citizen, and Grassroots Science: Towards a Transformative Scientific Research Model." In *Accountability Technologies—Tools for Asking Hard Questions*, ed. Dietmar Offenhuber and Katja Schechtner, 23–31. Vienna: Ambra V.

Douglas, Mary. 1966. *Purity and Danger; an Analysis of Concepts of Pollution and Taboo.* New York: Praeger.

Dourish, P. 2014. "NoSQL: The Shifting Materialities of Database Technology." *Computational Culture* 4.

Dourish, Paul, and Melissa Mazmanian. 2013. "Media as Material: Information Representations as Material Foundations for Organizational Practice." In *How Matter Matters: Objects, Artifacts, and Materiality in Organization Studies,* ed. Paul R. Carlile, Davide Nicolini, Ann Langley, and Haridimos Tsoukas, 92–118. Oxford: Oxford University Press.

Drechsler, Wolfgang. 2009. "The Rise and Demise of the New Public Management: Lessons and Opportunities for South East Europe." *Uprava-Administration* 7 (3): 7–27.

Drucker, Johanna. 2011. "Humanities Approaches to Graphical Display." *Digital Humanities Quarterly* 5 (1). http://www.digitalhumanities.org/dhq/vol/5/1/000091/000091.html.

Dunleavy, Patrick, Helen Margetts, Simon Bastow, and Jane Tinkler. 2006. "New Public Management Is Dead—Long Live Digital-Era Governance." *Journal of Public Administration: Research and Theory* 16 (3): 467–494.

Ebbesmeyer, Curtis C., W. James Ingraham, Thomas C. Royer, and Chester E. Grosch. 2007. "Tub Toys Orbit the Pacific Subarctic Gyre." *Eos, Transactions American Geophysical Union* 88 (1): 1–4. doi:10.1029/2007EO010001.

Edwards, Paul N. 2003. "Infrastructure and Modernity: Force, Time, and Social Organization in the History of Socio Technical Systems." In *Modernity and Technology,* ed. Thomas J. Misa, Philip Brey, and Andrew Feenberg, 185–225. Cambridge, MA: MIT Press.

Edwards, Paul N., Steven J. Jackson, Geoffrey C. Bowker, and Cory P. Knobel. 2007. "Understanding Infrastructure: Dynamics. Tensions. and Design." Working paper, final report for workshop, "History and Theory of Infrastructure: Lessons for New Scientific Cyberinfrastructures." Workshop at School of Information, University of Michigan.

Egyedi, Tineke M., and Donna C. Mehos. 2012. *Inverse Infrastructures: Disrupting Networks from Below.* Cheltenham, UK: Edward Elgar Publishing.

Ellard, Colin. 2009. *You Are Here: Why We Can Find Our Way to the Moon, but Get Lost in the Mall.* New York: Knopf Doubleday Publishing Group.

Elvidge, Christopher D., Kimberly E. Baugh, Paul C. Sutton, Budhendra Bhaduri, Benjamin T. Tuttle, Tilotamma Ghosh, Daniel Ziskin, and Edward H. Erwin. 2011. "Who's in the Dark—Satellite Based Estimates of Electrification Rates." In *Urban*

*Remote Sensing: Monitoring, Synthesis and Modeling in the Urban Environment*, ed. X. Yang, 211–224.Chichester, UK: John Wiley & Sons, Ltd.

Erickson, Thomas, and Wendy A. Kellogg. 2000. "Social Translucence: An Approach to Designing Systems That Support Social Processes." *ACM Transactions on Computer-Human Interaction* 7 (1): 59–83.

Estrella, Marisol, and John Gaventa. 1998. "Who Counts Reality?: Participatory Monitoring and Evaluation: A Literature Review." IDS Working Paper 70. Institute of Development Studies, Brighton, UK.

European Commission. 2008. "Directive 2008/98/EC on Waste (Waste Framework Directive)." http://eur-lex.europa.eu/legal-content/EN/TXT/?uri=CELEX:32008L0098.

Fahmi, Wael, and Keith Sutton. 2013. "Cairo's Contested Waste: The Zabaleen's Local Practices and Privatisation Policies." In *Organising Waste in the City: International Perspectives on Narratives and Practices*, ed. María José Zapata Campos and C. Michael Hall, 159–180. Policy Press at the University of Bristol. http://www.jstor.org/stable/j.ctt9qgpsv.15.

Fajnzylber, Pablo, William F. Maloney, and Gabriel V. Montes-Rojas. 2009. "Does Formality Improve Micro-Firm Performance? Quasi-Experimental Evidence from the Brazilian SIMPLES Program." IZA Discussion Paper No. 4531. https://papers.ssrn.com/sol3/papers.cfm?abstract_id=1501967.

Fathih, Mohamed Saif. 2015. "UNDP Launches App to Report Illegal Waste Dumping | Maldives Independent." December 21. http://maldivesindependent.com/environment/undp-launches-app-to-report-illegal-waste-dumping-120843.

Fay, Randy. 2012. "How Do Open Source Communities Govern Themselves?" *RandyFay.com*. March 5. http://randyfay.com/content/how-do-open-source-communities-govern-themselves.

FCC. 1997. "FCC CREATES NEW 311 CODE FOR NON-EMERGENCY POLICE CALLS AND 711 CODE FOR ACCESS TO TELECOMMUNICATIONS RELAY SERVICES." NEWSReport CC 97–7. http://transition.fcc.gov/Bureaus/Common_Carrier/News_Releases/1997/nrcc7014.txt.

Ferguson, Russell, Francis Alÿs, and Ann Philbin. 2008. *Francis Alÿs: The Politics of Rehearsal*. Göttingen: Steidl.

Fergutz, Oscar, Sonia Dias, and Diana Mitlin. 2011. "Developing Urban Waste Management in Brazil with Waste Picker Organizations." *Environment and Urbanization* 23 (2): 597–608. doi:10.1177/0956247811418742.

Floridi, Luciano. 2011. *The Philosophy of Information*. Oxford: Oxford University Press.

Flynn, Kevin. 2001. "20% Increase in 911 Calls Is Seen as a Result of Cellular Phone Use." *New York Times*, May 1. http://www.nytimes.com/2001/05/01/nyregion/2 0-increase-in-911-calls-is-seen-as-a-result-of-cellular-phone-use.html?src=pm.

Forrester, Jay W. 1970. "Systems Analysis as a Tool for Urban Planning." *IEEE Transactions on Systems Science and Cybernetics* 6 (4): 258–265. doi:10.1109/TSSC.1970 .300299.

Foucault, Michel. 1977. *Discipline and Punish: The Birth of the Prison.* New York: Pantheon Books.

Fox, Jonathan A. 2015. "Social Accountability: What Does the Evidence Really Say?" *World Development* 72 (August): 346–361. doi:10.1016/j.worlddev.2015.03.011.

Frischmann, Brett M. 2012. *Infrastructure: The Social Value of Shared Resources.* Oxford: Oxford University Press.

Fuller, Matthew, and Usman Haque. 2008. *Urban Versioning System 1.0.* New York: The Architectural League of New York.

Gabler, Ellen. 2010. "Those Pesky Bedbugs Won't Rest." *Chicago Tribune.* October 28. http://articles.chicagotribune.com/2010-10-28/a-z/ct-met-bedbugs-20101028_1 _bedbugs-pest-control-apartment-building-tenants.

Gabrys, Jennifer. 2013. *Digital Rubbish: A Natural History of Electronics.* Ann Arbor: University of Michigan Press.

Gaffin, Adam. 2011. "Saving Possums in the Dark." *Universal Hub.* February 16. http://www.universalhub.com/2011/bostons-high-tech-possum-saver.

Gaines, Linda, Anant Vyas, and John Anderson. 2006. "Estimation of Fuel Use by Idling Commercial Trucks." *Transportation Research Record: Journal of the Transportation Research Board* 1983 (1): 91–98. doi:10.3141/1983-13.

Galloway, A., J. Brucker-Cohen, L. Gaye, E. Goodman, and D. Hill. 2004. "Design for Hackability." In *Proceedings of the 5th Conference on Designing Interactive Systems: Processes, Practices, Methods, and Techniques*, 363–366.

Garrett, Theodore L., and the American Bar Association. 2004. *The RCRA Practice Manual.* Chicago: American Bar Association.

Gauch, Sarah. 2003. "Egypt Dumps 'Garbage People.'" *Christian Science Monitor*, January 6. http://www.csmonitor.com/2003/0106/p07s02-woaf.html.

Gaver, William, Jake Beaver, and Steve Benford. 2003. "Ambiguity as a Resource for Design." *Proceedings of the Conference on Human Factors in Computing Systems*, 233–240.

Gerxhani, Klarita. 2004. "The Informal Sector in Developed and Less Developed Countries: A Literature Survey." *Public Choice* 120 (3–4): 267–300.

Gibson, David V., George Kozmetsky, and Raymond W. Smilor. 1992. *The Technopolis Phenomenon: Smart Cities, Fast Systems, Global Networks*. Lanham, MD: Rowman & Littlefield.

Gibson, James J. 1979. *The Ecological Approach to Visual Perception*. Boston: Houghton Mifflin.

Gille, Zsuzsa. 2012. "From Risk to Waste: Global Food Waste Regimes." *Sociological Review* 60 (December): 27–46. doi:10.1111/1467-954X.12036.

Glaser, Barney G., and Anselm L. Strauss. 1967. *The Discovery of Grounded Theory: Strategies for Qualitative Research*. Piscataway, NJ: Transaction Publishers.

Goldsmith, Stephen, and Susan Crawford. 2014. *The Responsive City: Engaging Communities Through Data-Smart Governance*. 1st ed. San Francisco: Jossey-Bass.

Goodchild, M. F. 2007. "Citizens as Sensors: Web 2.0 and the Volunteering of Geographic Information." *GeoFocus (Editorial)* 7:8–10.

Goodspeed, Robert. 2015. "Smart Cities: Moving beyond Urban Cybernetics to Tackle Wicked Problems." *Cambridge Journal of Regions, Economy and Society* 8 (1): 79–92. doi:10.1093/cjres/rsu013.

Gordon, Eric, and Jessica Baldwin-Philippi. 2013. "Making a Habit Out of Engagement: How the Culture of Open Data Is Reframing Civic Life." In *Beyond Transparency*, ed. Brett Goldstein and Lauren Dyson, 139–150. San Francisco: Code for America Press.

Gordon, Eric, and Paul Mihailidis. 2016. *Civic Media: Technology, Design, Practice*. Cambridge, MA: MIT Press; http://civicmediaproject.org/works/civic-media-project/index.

Gouvêa, Carlos Portugal, and Ana Carolina Monguilod. 2014. "A Reading on Possible Legal Structures for Businesses Involving Catadores, Appropriate to Brazilian Reality." In *Design Waste & Dignity*, ed. Maria Cecília Loschiavo dos Santos, Stuart Walker, Sylmara Lopes, and Francelino Gonçalves Dias, 203–226. São Paulo, SP: Editora Olhares.

Graham, Stephen. 2005. "Software-Sorted Geographies." *Progress in Human Geography* 29 (5): 562–580.

Graham, Stephen. 2009. *Disrupted Cities: When Infrastructure Fails*. London: Routledge.

Graham, Stephen, and Simon Marvin. 1996. *Telecommunications and the City: Electronic Spaces, Urban Places*. Hove, UK: Psychology Press.

Graham, Stephen, and Simon Marvin. 2001. *Splintering Urbanism: Networked Infrastructures, Technological Mobilities and the Urban Condition*. London: Routledge.

Graham, Stephen, and Nigel Thrift. 2007. "Out of Order Understanding Repair and Maintenance." *Theory, Culture & Society* 24 (3): 1–25. doi:10.1177/0263276407075954.

Greenfield, Adam. 2013. *Against the Smart City. (The City Is Here for You to Use Book 1).* Kindle edition. New York: Do projects.

Greenpeace International. 2008. "Following the E-Waste Trail—UK to Nigeria." http://www.greenpeace.org/international/en/multimedia/multimedia-archive/Photo-Essays1/following-the-e-waste-trail/.

Guénard, Marion. 2013. "Cairo Puts Its Faith in Ragpickers to Manage the City's Waste Problem." *The Guardian*, November 19, sec. World news. https://www.theguardian.com/world/2013/nov/19/cairo-ragpickers-zabaleen-egypt-recycling.

Guerrero, Peter F. 1991. "Toxic Chemicals: EPA's Toxic Release Inventory Is Useful But Can Be Improved." RCED-91–121. http://www.gao.gov/products/RCED-91-121.

Guins, Raiford. 2014. *Landfill Legend: A BIT of Game After.* Cambridge, MA: MIT Press.

Hall, R Cargill. 2001. *A History of the Military Polar Orbiting Meteorological Satellite Program.* Chantilly, VA: National Reconnaissance Office.

Hamilton, James. 2005. *Regulation Through Revelation: The Origin, Politics, and Impacts of the Toxics Release Inventory Program.* Cambridge, UK: Cambridge University Press.

Hamilton, Kevin, Karrie Karahalios, Christian Sandvig, and Motahhare Eslami. 2014. "A Path to Understanding the Effects of Algorithm Awareness." In *CHI '14 Extended Abstracts on Human Factors in Computing Systems, 631–42. CHI EA '14.* New York: ACM; 10.1145/2559206.2578883.

Hannan, M. A., Md. Abdulla Al Mamun, Aini Hussain, Hassan Basri, and R. A. Begum. 2015. "A Review on Technologies and Their Usage in Solid Waste Monitoring and Management Systems: Issues and Challenges." *Waste Management* 43 (September): 509–523. doi:10.1016/j.wasman.2015.05.033.

Hansen, Tom Børsen. 2005. "Grassroots Science—an ISYP Ideal." *ISYP Journal on Science and World Affairs* 1 (1): 61–72.

Hargittai, Eszter. 2015. "Is Bigger Always Better? Potential Biases of Big Data Derived from Social Network Sites." *Annals of the American Academy of Political and Social Science* 659 (1): 63–76. doi:10.1177/0002716215570866.

Harless, William. 2013. "Ben Berkowitz's Formula: Spot a Problem, Map It, Fix It." *PBS NewsHour.* January 11. http://www.pbs.org/newshour/rundown/ben-berkowitz/.

Hawkins, Gay. 2005. *The Ethics of Waste: How We Relate to Rubbish.* Lanham, MD: Rowman & Littlefield.

Hearn, Mike. 2013. *Future of Money—Turing Festival 2013*. Edinburgh. https://www.youtube.com/watch?v=Pu4PAMFPo5Y.

Heatherley, Alan P. 2012. "Taxonomy/Classification of 311 Issues." *Open311 Discuss*. http://lists.open311.org/groups/discuss/messages/post/4bmfp2HaTclEL8fZvCmFJQ.

Heidegger, Martin. 1927. *Sein und Zeit*. Halle an der Saale: M. Niemeyer.

Heiman, Michael K. 1997. "Science by the People: Grassroots Environmental Monitoring and the Debate Over Scientific Expertise." *Journal of Planning Education and Research* 16 (4): 291–299. doi:10.1177/0739456X9701600405.

Hellawell, John M. 1991. "Development of a Rationale for Monitoring." In *Monitoring for Conservation and Ecology*, ed. F. B. Goldsmith, 1–14. Dordrecht: Springer. http://link.springer.com/chapter/10.1007/978-94-011-3086-8_1.

Hemingway, Ernest. 1932. *Death in the Afternoon*. New York: Charles Scribner's Sons.

Henderson, J. Vernon, Adam Storeygard, and David N. Weil. 2009. "Measuring Economic Growth from Outer Space." Working Paper 15199. National Bureau of Economic Research. http://www.nber.org/papers/w15199.

Henley, Andrew, G. Reza Arabsheibani, and Francisco G. Carneiro. 2009. "On Defining and Measuring the Informal Sector: Evidence from Brazil." *World Development* 37 (5): 992–1003.

Hering, Rudolph. 1912. "The Need for More Accurate Data in Refuse Disposal Work." *American Journal of Public Health* 2 (12): 909–911.

Hering, Rudolf, and Samuel A. Greeley. 1921. *Collection and Disposal of Municipal Refuse*. New York: McGraw-Hill Book Company, Inc.

Hill, Kashmir. 2016a. "Internet Mapping Turned a Remote Farm into a Digital Hell." *Fusion*. http://fusion.net/story/287592/internet-mapping-glitch-kansas-farm/.

Hill, Kashmir. 2016b. "Why Do People Keep Coming to This Couple's Home Looking for Lost Phones?" *Fusion*. http://fusion.net/story/214995/find-my-phone-apps-lead-to-wrong-home/.

Holovaty, Adrian. 2005. "Announcing Chicagocrime.org." *Holovaty.com*. May 18. http://www.holovaty.com/writing/chicagocrime.org-launch/.

Hood, Christopher. 1995. "The 'New Public Management' in the 1980s: Variations on a Theme." *Accounting, Organizations and Society* 20 (2–3): 93–109. doi:10.1016/0361-3682(93)E0001-W.

Hoornweg, Daniel, and Perinaz Bhada-Tata. 2012. "What a Waste: A Global Review of Solid Waste Management." *Urban Development Series Knowledge Papers* 15:1–98.

Hossain, Naomi. 2010. "Rude Accountability: Informal Pressures on Frontline Bureaucrats in Bangladesh." *Development and Change* 41 (5): 907–928. doi:10.1111/j.1467-7660.2010.01663.x.

Hughes, Thomas Parke. 1987. "The Evolution of Large Technological Systems." In *The Social Construction of Technological Systems*, ed. Wiebe E. Bijker, Thomas P. Hughes, and Trevor Pinch, 51–82. Cambridge, MA: MIT Press.

Hughes, Thomas Parke. 1998. *Rescuing Prometheus*. New York: Pantheon.

Hughes, Thomas Parke. 2004. *Human-Built World: How to Think about Technology and Culture*. Chicago: University of Chicago Press.

Ibem, Eziyi O. 2009. "Community-Led Infrastructure Provision in Low-Income Urban Communities in Developing Countries: A Study on Ohafia, Nigeria." *Cities* 26 (3): 125–132. doi:10.1016/j.cities.2009.02.001.

International Labour Office. 1972. *Employment, Incomes and Equality : A Strategy for Increasing Productive Employment in Kenya*. Geneva: International Labour Office.

Irwin, Alan. 1995. *Citizen Science: A Study of People, Expertise and Sustainable Development*. London; New York: Routledge.

ITCP-FGV. 2012. "Diagnóstico de Cooperativas de Catadores Coocares—PE." Abreu e Lima. http://www.cataacao.org.br/wp-content/uploads/2012/11/PNI-Coocares.pdf.

Jackson, Shannon. 2011. "High Maintenance: The Sanitation Aesthetics of Mierle Laderman Ukeles." In *Social Works: Performing Art, Supporting Publics*, 75–104. Taylor & Francis.

Jacobs, Jane. 1961. *The Death and Life of Great American Cities*. New York: Random House.

Jensen, Casper Bruun, and Brit Ross Winthereik. 2013. *Monitoring Movements in Development Aid: Recursive Partnerships and Infrastructures*. Cambridge, MA: MIT Press.

Jimenez, Alberto Escalada, Adrian Dabrowski, Noburu Sonehara, Juan M. Montero Martinez, and Isao Echizen. 2014. "Tag Detection for Preventing Unauthorized Face Image Processing." In *Proceedings of the 13th International Workshop on Digital-Forensics and Watermarking*, 1–12.

Johnson, Steven. 2010. "What a Hundred Million Calls to 311 Reveal About New York," November 1. https://www.wired.com/magazine/2010/11/ff_311_new_york/all/1.

Jones, Peter Tom, Daneel Geysen, Yves Tielemans, Steven Van Passel, Yiannis Pontikes, Bart Blanpain, Mieke Quaghebeur, and Nanne Hoekstra. 2013. "Enhanced Landfill Mining in View of Multiple Resource Recovery: A Critical Review." *Journal of Cleaner Production*. Special Volume: Urban and Landfill Mining 55 (September): 45–55. doi:10.1016/j.jclepro.2012.05.021.

Joshi, Anuradha, and Peter Houtzager. 2012. "Widgets or Watchdogs? Conceptual Explorations in Social Accountability." *Public Management Review* 14 (2): 145–162.

Kang, Hai-Yong, and Julie M. Schoenung. 2005. "Electronic Waste Recycling: A Review of U.S. Infrastructure and Technology Options." *Resources, Conservation and Recycling* 45 (4): 368–400. doi:10.1016/j.resconrec.2005.06.001.

Kelty, Christopher. 2005. "Geeks, Social Imaginaries, and Recursive Publics." *Cultural Anthropology* 20 (2): 185–214. doi:10.1525/can.2005.20.2.185.

Kempton, Willett. 1986. "Two Theories of Home Heat Control." *Cognitive Science* 10 (1): 75–90.

Kensing, F., and J. Blomberg. 1998. "Participatory Design: Issues and Concerns." *Computer Supported Cooperative Work* 7 (3): 167–185.

King, Stephen F., and Paul Brown. 2007. "Fix My Street or Else: Using the Internet to Voice Local Public Service Concerns." In *Proceedings of the 1st International Conference on Theory and Practice of Electronic Governance*, ed. Tomasz Janowski and Theresa A. Pardo, 72–80. ICEGOV '07. New York: ACM. doi:10.1145/1328057.1328076.

Kirschenbaum, Matthew G. 2008. *Mechanisms: New Media and the Forensic Imagination*. Cambridge, MA: MIT Press.

Kitchin, Rob. 2014. *The Data Revolution: Big Data, Open Data, Data Infrastructures and Their Consequences*. Thousand Oaks, CA: SAGE Publications Ltd.

Kitchin, Rob, and Martin Dodge. 2011. *Code/Space: Software and Everyday Life*. Cambridge, MA: MIT Press.

Kramer, Adam D. I. 2012. "The Spread of Emotion via Facebook." In *Proceedings of the SIGCHI Conference on Human Factors in Computing Systems*, 767–770. CHI '12. New York: ACM. doi:10.1145/2207676.2207787.

Kreith, Frank, and George Tchobanoglous. 2002. *Handbook of Solid Waste Management*. New York: McGraw-Hill Professional.

Kroah-Hartman, Greg, Jonathan Corbet, and Amanda McPherson. 2008. "Linux Kernel Development." *The Linux Foundation*. http://www.linuxfoundation.org/sites/main/files/publications/linuxkerneldevelopment.pdf.

Kummer, Katharina. 1999. *International Management of Hazardous Wastes: The Basel Convention and Related Legal Rules*. Oxford, UK: Oxford University Press.

Kuznetsov, Stacey, and Eric Paulos. 2010. "Rise of the Expert Amateur: DIY Projects, Communities, and Cultures." In *Proceedings of the 6th Nordic Conference on Human-Computer Interaction: Extending Boundaries*, 295–304. http://dl.acm.org/citation.cfm?id=1868950.

Landemore, Hélène. 2015. "Inclusive Constitution-Making: The Icelandic Experiment." *Journal of Political Philosophy* 23 (2): 166–191. doi:10.1111/jopp.12032.

Lathrop, Daniel, and Laurel Ruma. 2010. *Open Government.* Sebastopol, CA: O'Reilly Media, Inc.

Latour, Bruno. 1994. On Technical Mediation. *Common Knowledge* 3 (2): 29–64.

Latour, Bruno. 1999. *Pandora's Hope: Essays on the Reality of Science Studies.* Cambridge, MA: Harvard University Press.

Latour, Bruno, and Emile Hermant. 2004. *Paris: Invisible City.* http://www.brunolatour.fr/virtual/index.html#.

Lave, Jean, and Etienne Wenger. 1991. *Situated Learning : Legitimate Peripheral Participation.* Cambridge, UK; New York: Cambridge University Press.

Lazer, David, Ryan Kennedy, Gary King, and Alessandro Vespignani. 2014. "The Parable of Google Flu: Traps in Big Data Analysis." *Science* 343 (6176): 1203–1205. doi:10.1126/science.1248506.

Lee, David, Dietmar Offenhuber, Assaf Biderman, and Carlo Ratti. 2014. "Learning from Tracking Waste: How Transparent Trash Networks Affect Sustainable Attitudes and Behavior." In *World Forum on Internet of Things (WF-IoT),* 2014 IEEE, 130–134.

Lee, J. A., and V. M. Thomas. 2004. "GPS and Radio Tracking of End-of-Life Products [Recycling and Waste Disposal Applications]." In *Proceedings of the International Symposium on Electronics and the Environment,* 309–312. IEEE.

Lepawsky, Josh. 2014. "The Changing Geography of Global Trade in Electronic Discards: Time to Rethink the E-Waste Problem." *The Geographical Journal* 181 (2): 147–159. doi:10.1111/geoj.12077.

Lepawsky, Josh, and Chris Mcnabb. 2010. "Mapping International Flows of Electronic Waste." *Canadian Geographer* 54 (2): 177–195. doi:10.1111/j.1541-0064. 2009.00279.x.

Lessig, Lawrence. 2009. "Against Transparency." *The New Republic,* October 9. http://www.tnr.com/print/article/books-and-arts/against-transparency.

Lewis, Herbert S. 1993. "A New Look at Actor-Oriented Theory." *Political and Legal Anthropology Review* 16 (3): 49–56.

Liboiron, Max. 2009. "Recycling as a Crisis of Meaning." *eTopia.* http://etopia .journals.yorku.ca/index.php/etopia/article/view/36718.

Lieberman, J. Ben. 1967. *Types of Typefaces and How to Recognize Them.* New York: Sterling Pub. Co.

Lindblom, Charles E. 1959. "The Science of 'Muddling through.'" *Public Administration Review* 19 (2): 79–88.

Long, Norman. 1989. *Encounters at the Interface: A Perspective on Social Discontinuities in Rural Development.* Wageningen Studie in Sociology 27. Wageningen, NL: Pudoc.

Loukissas, Yanni Alexander. 2017. "Taking Big Data Apart: Local Readings of Composite Media Collections." *Information Communication and Society* 20 (5): 1–14. doi:10.1080/1369118X.2016.1211722.

Lynch, Kevin. 1955. "Progress Report and Plan for Future Studies—June 1955." Massachusetts Institute of Technology, Institute Archives and Special Collections. http://dome.mit.edu/handle/1721.3/35707.

Lynch, Kevin. 1960. *The Image of the City.* Cambridge, MA: MIT Press.

Lynch, Kevin. 1972. *What Time Is This Place?* Cambridge, MA: MIT Press.

Lynch, Kevin. 1984. *Good City Form.* Cambridge, MA: MIT Press.

Lynch, Kevin, and Gary Hack. 1971. *Site Planning.* Cambridge, MA: MIT Press.

Lynch, Kevin, with Michael Southworth, ed. 1991. *Wasting Away—an Exploration of Waste: What It Is, How It Happens, Why We Fear It, How to Do It Well.* New York: Random House, Inc.

MacBride, Samantha. 2012. *Recycling Reconsidered: The Present Failure and Future Promise of Environmental Action in the United States.* Cambridge, MA: MIT Press.

Macedo, Ana Patrícia de Aguiar Teixeira, and Ricardo Cavalcanti Furtado. 2003. "Uma Avaliação de Desempenho Na Coleta E Transporte Da Gestão de Resíduos Sólidos Domiciliares Nos Municípios de Recife, Olinda E Jaboatão Dos Guararapes." http://www.abepro.org.br/biblioteca/ENEGEP2003_TR1004_0249.pdf.

MacKinnon, Rebecca. 2012. *Consent of the Networked: The Worldwide Struggle For Internet Freedom.* First Trade Paper Edition. New York: Basic Books.

Maisonneuve, Nicolas, Matthias Stevens, Maria E. Niessen, and Luc Steels. 2009. "NoiseTube: Measuring and Mapping Noise Pollution with Mobile Phones." In *Information Technologies in Environmental Engineering,* ed. Ioannis N. Athanasiadis, Pericles A. Mitkas, Andrea E. Rizzoli, and Jorge Marx Gómez, 215–228. Berlin: Springer Science & Business Media.

Maloney, William F. 2004. "Informality Revisited." *World Development* 32 (7): 1159–1178. doi:10.1016/j.worlddev.2004.01.008.

Maniates, M. F. 2001. "Individualization: Plant a Tree, Buy a Bike, Save the World?' *Global Environmental Politics* 1 (3): 31–52.

Mantelero, Alessandro. 2013. "The EU Proposal for a General Data Protection Regulation and the Roots of the 'Right to Be Forgotten.'" *Computer Law & Security Review* 29 (3): 229–235.

Marques, Rui Cunha, and Nuno Ferreira da Cruz. 2015. *Recycling and Extended Producer Responsibility: The European Experience*. Farnham, UK: Ashgate Publishing, Ltd.

Martínez, J. A., K. H. Pfeffer, and Tara van Dijk. 2009. "The Capacity of E-Government Tools: Claimed Potentials, Unnamed Limitations." In *Proceedings of the 10th N-AERUS Conference: Challenges to Open Cities in Africa, Asia, Latin America and the Middle East: Shared Spaces within and Beyond, IHS Rotterdam.*

Mayer-Schönberger, Viktor, and Kenneth Cukier. 2013. *Big Data: A Revolution That Will Transform How We Live, Work, and Think*. Boston: Eamon Dolan/Houghton Mifflin Harcourt.

Mayntz, Renate, and Thomas Parke Hughes. 1988. *The Development of Large Technical Systems*. Boulder, CO: Westview Press.

Mazerolle, Lorraine, Dennis Rogan, James Frank, Christine Famega, and John E. Eck. 2002. "Managing Citizen Calls to the Police: The Impact of Baltimore's 3-1-1 Call System." *Criminology & Public Policy* 2 (1): 97–124. doi:10.1111/j.1745-9133.2002. tb00110.x.

Medina, Martin. 2007a. *The World's Scavengers: Salvaging for Sustainable Consumption and Production*. Lanham: AltaMira Press.

Medina, Martin. 2007b. "Waste Picker Cooperatives in Developing Countries." In *Membership-Based Organizations of the Poor*, ed. Martha Alter Chen, 105–121. Abingdon, UK: Routledge.

Medina, Martin. 2008. "The Informal Recycling Sector in Developing Countries: Organizing Waste Pickers to Enhance Their Impact," October. https://openknowledge.worldbank.org/handle/10986/10586.

Medina, Martin. 2010. "World's Largest And Most Dynamic Scavenger Movement." *BioCycle* 51 (10): 32–33.

Meier, Patrick. 2015. *Digital Humanitarians: How Big Data Is Changing the Face of Humanitarian Response*. Boca Raton, FL: CRC Press.

Meier, Patrick, and Jennifer Leaning. 2009. "Applying Technology to Crisis Mapping and Early Warning in Humanitarian Settings." Crisis Mapping Working Paper I of III. Harvard Humanitarian Initiative.

Ministério Público de Pernambuco. 2009. "Catadores Dão Grito Pela Sobrevivência." June 17. http://www.mppe.mp.br/siteantigo/siteantigo.mppe.mp.br/index.pl/clipagem20091706_catadores.html.

Melosi, Martin V. 2004. *Garbage in the Cities: Refuse Reform and the Environment*. Pittsburgh, PA: University of Pittsburgh Press.

Michaelis, Peter. 1995. "Product Stewardship, Waste Minimization and Economic Efficiency: Lessons from Germany." *Journal of Environmental Planning and Management* 38 (2): 231–244. doi:10.1080/09640569513039.

Migliano, João Ernesto Brasil, Jacques Demajorovic, and Lucia Helena Xavier. 2014. "Shared Responsibility and Reverse Logistics Systems for E-Waste in Brazil." *Journal of Operations and Supply Chain Management* 7 (2): 91–109.

Ministerio do Meio Ambiente. 2010. "Legislação Brasileira Prevê Fim Dos Lixões." July 30. http://www.mma.gov.br/informma/item/6462-legislacao-brasileira-preve-fim-dos-lixoes.

Mitchell, William J. 1995. *City of Bits: Space, Place, and the Infobahn.* Cambridge, MA: MIT Press.

Mongkol, Kulachet. 2011. "The Critical Review of New Public Management Model and Its Criticisms." *Research Journal of Business Management* 5 (1): 35–43. doi:10.3923/rjbm.2011.35.43.

Moore, Sarah A. 2008. "The Politics of Garbage in Oaxaca, Mexico." *Society & Natural Resources* 21 (7): 597–610.

Morozov, Evgeny. 2014. *To Save Everything, Click Here: The Folly of Technological Solutionism.* New York: PublicAffairs.

Morris, Jeffrey. 2005. "Comparative LCAs for Curbside Recycling Versus Either Landfilling or Incineration with Energy Recovery." *International Journal of Life Cycle Assessment* 10 (4): 273–284. doi:10.1065/lca2004.09.180.10.

Mudge, Stephen M. 2008. "Environmental Forensics and the Importance of Source Identification." In *Issues in Environmental Science and Technology*, ed. R. E. Hester and R. M. Harrison, 1–16. Cambridge, UK: Royal Society of Chemistry.

Nagle, Robin. 2013. *Picking Up: On the Streets and Behind the Trucks with the Sanitation Workers of New York City.* New York: Farrar, Straus and Giroux.

Mechanics, New Urban. 2014. "City Hall To Go." http://newurbanmechanics.org/project/city-hall-to-go/.

New York City. 2013. "Mayor Bloomberg Commemorates Ten Years of NYC311, the Nation's Largest and Most Comprehensive 311 Service | City of New York." March 11. http://www1.nyc.gov/office-of-the-mayor/news/089-13/mayor-bloomberg-commemorates-ten-years-nyc311-nation-s-largest-most-comprehensive-311.

Nielsen, Jacob. 2006. "Participation Inequality: The 90-9-1 Rule for Social Features." October 9. https://www.nngroup.com/articles/participation-inequality/.

Norman, Donald A. 2002. *The Design of Everyday Things.* New York: Basic Books.

Norman, Donald A., and Pieter Jan Stappers. 2015. "DesignX: Complex Sociotechnical Systems." *She Ji: The Journal of Design, Economics, and Innovation* 1 (2): 83–106.

Norton-Arnold & Co., URS Corp., and Herrera, Inc. 2007. "Seattle Solid Waste Recycling, Waste Reduction, and Facilities Opportunities, Volume 2." https://www.seattle.gov/util/cs/groups/public/@spu/@garbage/documents/webcontent/spu01_002547.pdf.

Nye, David E. 1992. *Electrifying America: Social Meanings of a New Technology, 1880–1940*. Cambridge, MA: The MIT Press.

Obama, Barack. 2009. "Transparency and Open Government." January 29. Memorandum for the Heads of Executive Departments and Agencies. Executive Office of the President. http://web.archive.org/web/20090130015956/http://www.whitehouse.gov/the_press_office/TransparencyandOpenGovernment.

O'Brien, Daniel Tumminelli, Robert J. Sampson, and Christopher Winship. 2015. "Ecometrics in the Age of Big Data: Measuring and Assessing 'Broken Windows' Using Large-Scale Administrative Records." *Sociological Methodology* 45 (1): 101–147. doi:10.1177/0081175015576601.

O'Connell, Pamela Licalzi. 2005. "Do-It-Yourself Cartography." *The New York Times Magazine*, December 11. http://www.nytimes.com/2005/12/11/magazine/11ideas1-13.html.

OECD. 2001. *Extended Producer Responsibility*. Paris: Organisation for Economic Co-operation and Development. http://www.oecd-ilibrary.org/content/book/9789264189867-en.

OED. 2012. "'Infrastructure, N.'" Definition. *Oxford English Dictionary*. http://www.oed.com/view/Entry/95624.

Offenhuber, Dietmar. 2008. "The Invisible Display—Design Strategies for Ambient Media in the Urban Context." In *Proceedings of the 2nd Workshop on Ambient Information Systems*. Colocated with Ubicomp. Seoul, Korea.

Offenhuber, Dietmar. 2014. "The Designer as Regulator—Design Patterns and Categorization in Citizen Feedback Systems." In *Proceedings of the 2014 Workshop on Big Data and Urban Informatics*. Chicago.

Offenhuber, Dietmar. 2015. "Infrastructure Legibility—a Comparative Analysis of open311-Based Citizen Feedback Systems." *Cambridge Journal of Regions, Economy and Society* 8 (1): 93–112. doi:10.1093/cjres/rsu001.

Offenhuber, D., and D. Lee. 2012. "Putting the Informal on the Map: Tools for Participatory Waste Management." In *Proceedings of the 12th Participatory Design Conference: Exploratory Papers, Workshop Descriptions, Industry Cases—Volume 2*, 13–16. http://dl.acm.org/citation.cfm?id=2348144.2348150.

Offenhuber, Dietmar, and Katja Schechtner, eds. 2013. *Accountability Technologies—Tools for Asking Hard Questions*. Vienna: Ambra V.

Offenhuber, Dietmar, Malima I. Wolf, and Carlo Ratti. 2013. "Trash Track—Active Location Sensing for Evaluating E-Waste Transportation." *Waste Management & Research* 31 (2): 150–159. doi:10.1177/0734242X12469822.

Offenhuber, Dietmar, David Lee, Malima I. Wolf, Santi Phithakkitnukoon, Assaf Biderman, and Carlo Ratti. 2012. "Putting Matter in Place." *Journal of the American Planning Association* 78 (2): 173–196. doi:10.1080/01944363.2012.677120.

Okolloh, Ory. 2009. "Ushahidi, or 'Testimony': Web 2.0 Tools for Crowdsourcing Crisis Information." *Participatory Learning and Action* 59 (June): 65–70.

O'Reilly, Tim. 2011. "Government as a Platform." *Innovations: Technology, Governance, Globalization* 6 (1): 13–40. doi:10.1162/INOV_a_00056.

Osborne, David, and Ted Gaebler. 1992. *Reinventing Government: How the Entrepreneurial Spirit Is Transforming the Public Sector*. New York: Plume.

Ostrom, Elinor. 1996. "Crossing the Great Divide: Coproduction, Synergy, and Development." *World Development* 24 (6): 1073–1087. doi:10.1016/0305-750X(96)00023-X.

Pachube. 2008. *Archive.org*. May 5. http://web.archive.org/web/20080505045408/http://pachube.com/.

Pardo, José Luis. 2006. "Nunca Fue Tan Hermosa La Basura (Never Was Trash so Beautiful)." In *Distorsiones Urbanas (Urban Distortions)*, ed. Basurama, 66–76. Madrid: Basurama. http://www.basurama.org/b06_distorsiones_urbanas_pardo_e.htm.

Parkman, E. Breck. 2014. "A Hippie Discography: Vinyl Records from a Sixties Commune." *World Archaeology* 46 (3): 431–447.

Pask, Gordon. 1976. *Conversation Theory: Applications in Education and Epistemology*. Amsterdam; New York: Elsevier.

Peattie, Lisa Redfield. 1987. *Planning, Rethinking Ciudad Guayana*. Ann Arbor: University of Michigan Press.

Peirce Edition Project, ed. 1998. *The Essential Peirce, Volume 2: Selected Philosophical Writings, 1893-1913*, 4–11. Bloomington: Indiana University Press.

Pellow, D. N. 2004. *Garbage Wars: The Struggle for Environmental Justice in Chicago*. Cambridge, MA: MIT Press.

Phithakkitnukoon, Santi, Malima I. Wolf, Dietmar Offenhuber, David Lee, Assaf Biderman, and Carlo Ratti. 2013. "Tracking Trash." *IEEE Pervasive Computing* 12 (2): 38–48. doi:10.1109/MPRV.2013.37.

Pollitt, Christopher, and Geert Bouckaert. 2004. *Public Management Reform: A Comparative Analysis*. New York: Oxford University Press.

Porter, Richard C. 2002. *The Economics of Waste*. Washington, DC: Resources for the Future.

Powell, Jerry. 2013. "Abandoned Warehouses Full of CRTs Found in Several States." *Resource Recycling*. https://resource-recycling.com/e-scrap/2013/08/23/abandoned -warehouses-full-crts-found-several-states.

Prefeitura de Olinda. 2010. "Sítio Histórico de Olinda Terá Coleta Seletiva de Lixo." *Prefeitura de Olinda*. December 23. http://www.olinda.pe.gov.br/servicos-publicos/ sitio-historico-de-olinda-tera-coleta-seletiva-de-lixo.

Prud'homme, Remy. 2005. "Infrastructure and Development." In *Annual World Bank Conference on Development Economics 2005: Lessons of Experience*, ed. Francois Bourguignon and Boris Pleskovic, 153–192. Washington, DC: World Bank Publications.

Puckett, Jim, Sarah Westerveldt, Richard Gutierrez, and Yuka Takamiya. 2005. "The Digital Dump: Exporting Re-Use and Abuse to Africa." Seattle, WA. http:// archive.ban.org/library/TheDigitalDump.pdf.

Puckett, Jim, M. Ryan, B. Zude, S. Bernstein, T. Hedenrick, and Basel Action Network. 2002. "Exporting Harm: The High-Tech Trashing of Asia." Seattle: Basel Action Network.

Rafaeli, Sheizaf, and Yaron Ariel. 2008. "Online Motivational Factors: Incentives for Participation and Contribution in Wikipedia." In *Psychological Aspects of Cyberspace: Theory, Research, Applications*, ed. A. Barak, 243–267. New York: Cambridge University Press.

Rāmasvāmi, Gītā. 2005. *India Stinking: Manual Scavengers in Andhra Pradesh and Their Work*. Pondicherry: Navayana.

Rams, Dieter. 1984. "Omit the Unimportant." *Design Issues* 1 (1):24–26.

Rathje, William, and Cullen Murphy. 2001. *Rubbish! The Archaeology of Garbage*. Tucson, AZ: University of Arizona Press.

Ratto, Matt. 2014. "Critical Making." In *Open Design Now: Why Design Cannot Remain Exclusive*, ed. Bas van Abel, Lucas Evers, Peter Troxler and Roel Klaassen, 202–214. Amsterdam, NL: BIS Publishers.

Ratto, Matt, Megan Boler, and Ronald Deibert. 2014. *DIY Citizenship: Critical Making and Social Media*. Cambridge, MA: MIT Press.

Raymond, Eric. 1999. "The Cathedral and the Bazaar." *Knowledge, Technology & Policy* 12 (3): 23–49. doi:10.1007/s12130-999-1026-0.

RecycleNet Corporation. 2010. "ScrapIndex.com." http://www.scrapindex.com/spotmarket/.

Recycling Today. 2013. "China's 'Green Fence' Continues to Clog Export Markets." *Recycling Today*. March 21. http://www.recyclingtoday.com/article/europe-inspections-china-customs.

Repa, Edward. 2005. "Interstate Movement of Municipal Solid Waste." 05–2. NSWMA Research Bulletin. http://environmentalisteveryday.org/docs/research-bulletin/Research-Bulletin-Interstate-Waste-2005.pdf.

Riegler, Alexander. 2007. "Superstition in the Machine." In *Anticipatory Behavior in Adaptive Learning Systems*, ed. Martin V. Butz, Olivier Sigaud, Giovanni Pezzulo, and Gianluca Baldassarre, 4520:57–72. Berlin, Heidelberg: Springer.

Roth, Matthew. 2009. "SeeClickFix: Is 'Little Brother' the Next Big Thing?" *Streetsblog New York City*. March 25. http://nyc.streetsblog.org/2009/03/25/seeclickfix-is-little-brother-the-next-big-thing/.

Royte, Elizabeth. 2005. *Garbage Land: On the Secret Trail of Trash*. Boston: Little, Brown and Company.

Saar, S., and V. Thomas. 2002. "Toward Trash That Thinks: Product Tags for Environmental Management." *Journal of Industrial Ecology* 6 (2): 133–146.

Santos, Maria Cecília Loschiavo dos, Stuart Walker, Sylmara Lopes, and Francelino Gonçalves Dias, eds. 2014. *Design Waste & Dignity*. São Paulo, SP: Editora Olhares.

Sassen, Saskia. 2011. "Open Source Urbanism." *Domus*, June 29. http://www.domusweb.it/en/op-ed/2011/06/29/open-source-urbanism.html.

Scartezini, Vanda. 2013. "Redeeming E-Waste in Brazil." *Waste Management World*, April 1. https://waste-management-world.com/a/redeeming-e-waste-in-brazil.

Scharfenberg, Jeremy, Lauren Pederson, and Anne Choate. 2004. "WARM Model Transportation Research—Draft." Methane Contract 009, Task 06. ICF International. http://www.epa.gov/climatechange/wycd/waste/SWMGHreport.html.

Schedler, Andreas. 1999. "Conceptualizing Accountability." In *The Self-Restraining State: Power and Accountability in New Democracies*, ed. Andreas Schedler, Larry Diamond, and Marc F. Plattner, 13–28. Boulder, CO: Lynne Rienner Publishers Inc.

Scheinberg, Anne. 2010. "Improving Recycling Rates through Privatization and Economic Incentives to Informal Sector." Presented at the Second Meeting of the Regional 3R Forum in Asia, Kuala Lumpur, Malaysia, October. http://www.uncrd.or.jp/env/3r_02/.

Scheinberg, Anne. 2012. "Informal Sector Integration and High Performance Recycling: Evidence from 20 Cities." WIEGO Working Paper (Urban Policies) No. 23. Manchester: Women in Informal Employment Globalizing and Organizing

(WIEGO). http://www.wiego.org/sites/default/files/publications/files/Scheinberg _WIEGO_WP23.pdf.

Scheinberg, Anne, David C. Wilson, and Ljiljana Rodic. 2010. *Solid Waste Management in the World's Cities: Water and Sanitation in the World's Cities 2010.* London; Washington, DC: UN-HABITAT/Earthscan.

Scheinberg, Anne, Jelena Nesić, Rachel Savain, Pietro Luppi, Portia Sinnott, Flaviu Petean, and Flaviu Pop. 2016. "From Collision to Collaboration–Integrating Informal Recyclers and Re-Use Operators in Europe: A Review." *Waste Management & Research* 34 (9): 820–839.

Schmidt, Benjamin M. 2013. "Words Alone: Dismantling Topic Models in the Humanities." *Journal of Digital Humanities* 2 (1): 49–65.

Schubeler, Peter. 1996. *Participation and Partnership in Urban Infrastructure Management.* Washington DC: World Bank Publications.

Schudson, Michael. 1998. "Changing Concepts of Democracy." In MIT *Communications Forum.* Cambridge, MA. http://web.mit.edu/comm-forum/papers/schudson .html.

Scott, James C. 1999. *Seeing Like a State: How Certain Schemes to Improve the Human Condition Have Failed.* New Haven, CT: Yale University Press.

Seattle Public Utilities. 2003. "Destinations and End Used of Materials Collected by Seattle's Curbside Recycling Program." https://www.seattle.gov/util/cs/groups/ public/@spu/@recycle/documents/webcontent/endproduc_200312021144134.pdf.

Seattle Public Utilities. 2010. "Seattle Public Utilities—Contracts." http://www.seattle .gov/util/myservices/garbage/aboutgarbage/contracts/.

SeeClickFix. 2007. "First Jam Session." *SeeClickFix.* December 2. https://gov .seeclickfix.com/2007/12/02/first-jam-session/.

SeeClickFix. 2011. "SeeClickFix Integrates with Boston Open 311." *SeeClickFix.* September 12. https:/gov.seeclickfix.com/2011/09/12/seeclickfix-integrates-with -boston-open-311.

Senado Federal. 2012. "Vetada regulamentação de catador e reciclador de papel." Materia Agencia. October 1. http://www12.senado.gov.br/noticias/materias/2012/ 01/10/vetada-regulamentacao-de-catador-e-reciclador-de-papel.

Sennett, Richard. 1970. *The Uses of Disorder: Personal Identity and City Life.* New York: Knopf.

Senseable City Lab. 2009. "Trash | Track." http://senseable.mit.edu/trashtrack/.

Shah, Rajiv, Jay Kesan, and Andrew Kennis. 2008. "Implementing Open Standards: A Case Study of the Massachusetts Open Formats Policy." In *Proceedings of the 2008*

*International Conference on Digital Government Research*, 262–271. Digital Government Society of North America.

Shamiyeh, Michael. 2014. *Driving Desired Futures*. Basel, Switzerland: Birkhäuser.

Shaw Environmental, Inc. 2013. "An Analysis of the Demand for CRT Glass Processing in the U.S." White paper. Shaw Environmental, Inc., Baton Rouge, LA.

Shepard, Mark. 2011. *Sentient City: Ubiquitous Computing, Architecture, and the Future of Urban Space*. Cambridge, MA: MIT Press.

Short, John, Ederyn Williams, and Bruce Christie. 1976. *The Social Psychology of Telecommunications*. Hoboken, NJ: John Wiley & Sons.

Shulman, Robin. 2005. "A Man With a Vision for Getting New York Wired." *The New York Times*, September 2, sec. New York Region / Metro Campaigns. http://www.nytimes.com/2005/09/02/nyregion/metrocampaigns/a-man-with-a-vision-for-getting-new-york-wired.html.

Simpson-Herbert, Mayling. 2005. *A Paper Life: Belgrade's Roma in the Underworld of Waste Scavenging and Recycling*. Leicestershire, UK: Water, Engineering and Development Centre (WEDC), Loughborough University.

Singel, Ryan. 2005. "Map Hacks on Crack." *WIRED*, July 2. https://www.wired.com/2005/07/map-hacks-on-crack/.

Smith, Ted, David Allan Sonnenfeld, and David N. Pellow. 2006. *Challenging the Chip: Labor Rights and Environmental Justice in the Global Electronics Industry*. Philadelphia, PA: Temple University Press.

Snowcrew. 2012. "SnowCrew: Shoveling assistance for the elderly and people with disabilities." *Neighbors for Neighbors*. http://neighborsforneighbors.org/page/snowcrew.

Sparrow, Malcolm K., Mark H. Moore, and David M. Kennedy. 1992. *Beyond 911: A New Era for Policing*. New York: Basic Books.

Star, Susan Leigh. 1999. "The Ethnography of Infrastructure." *American Behavioral Scientist* 43 (3): 377–391. doi:10.1177/00027649921955326.

Star, Susan Leigh, and Karen Ruhleder. 1994. "Steps towards an Ecology of Infrastructure." In *Proceedings of the 1994 ACM Conference on Computer Supported Cooperative Work—CSCW '94*, 253–264. Chapel Hill, NC: ACM. doi:10.1145/192844.193021.

Sterling, B. 2004. "When Blobjects Rule the Earth." Keynote at SIGGRAPH 2004, August 9, Los Angeles.

Strasser, Susan. 1999. *Waste and Want: A Social History of Trash*. New York: Metropolitan Books.

Streicher-Porte, Martin. 2009. "Case Study Brazil." *Ewasteguide.info.* http://ewasteguide.info/node/4183.

Sutton, Paul. 1997. "Modeling Population Density with Night-Time Satellite Imagery and GIS." *Computers, Environment and Urban Systems* 21 (3): 227–244.

Swartz, Aaron. 2010. "When Is Transparency Useful?" *Open Government: Collaboration, Transparency, and Participation in Practice* 1:267–272.

Sweeney, Latanya. 2013. "Discrimination in Online Ad Delivery." *Queue* 11 (3): 10.

Tarr, Joel A. 1996. *The Search for the Ultimate Sink: Urban Pollution in Historical Perspective.* 1st ed. Akron, OH: University of Akron Press.

Tatum, Jesse S. 1992. "The Home-Power Movement and the Assumptions of Energy-Policy Analysis." *Energy* 17 (2): 99–107. doi:10.1016/0360-5442(92)90060-D.

Thaler, Richard H., and Cass R. Sunstein. 2008. *Nudge: Improving Decisions about Health, Wealth and Happiness.* London: Penguin Books.

Thompson, Michael. 1979. *Rubbish Theory: The Creation and Destruction of Value.* Oxford: Oxford University Press.

Tobler, W. R. 1970. "A Computer Movie Simulating Urban Growth in the Detroit Region." *Economic Geography* 46 (June): 234–240. doi:10.2307/143141.

Torrance, Morag. 2009. "Reconceptualizing Urban Governance through a New Paradigm for Urban Infrastructure Networks." *Journal of Economic Geography* 9 (6): 805–822.

Townsend, Anthony M. 2013. *Smart Cities: Big Data, Civic Hackers, and the Quest for a New Utopia.* New York: W.W. Norton & Company.

Toyama, Kentaro. 2015. *Geek Heresy: Rescuing Social Change from the Cult of Technology.* New York: PublicAffairs.

TransparentPlanet. 2012. *U.S. CRT Glass Management.* Bainbridge Island, WA: TransparentPlanet LLC.

Trumbull, Deborah J., Rick Bonney, Derek Bascom, and Anna Cabral. 2000. "Thinking Scientifically during Participation in a Citizen-Science Project." *Science Education* 84 (2): 265–275.

United Church of Christ. 1987. *Toxic Wastes and Race in the United States.* New York: Commission for Racial Justice of the United Church of Christ.

U.S. Congress. 1976. *Resource Conservation and Recovery Act. 42 U.S.C.* Vol. 6901.

U.S. Congress. 1984. *Resource Conservation and Recovery Act, Hazardous and Solid Wastes Amendments. 42 U.S.C.* Vol. 6901.

U.S. EPA. 2006. "Solid Waste Management and Greenhouse Gases: A Life-Cycle Assessment of Emissions and Sinks, 3rd Edition." https://www3.epa.gov/warm/ SWMGHGreport.htm.

U.S. EPA. 2009. "EPA Facility Registry Service (FRS)." December 7. http://www.epa .gov/enviro/html/fii/index.html.

U.S. EPA and Office of Air and Radiation. 2006. "Waste Reduction Model (WARM)." October 19. https://www.epa.gov/warm.

U.S. EPA Office of Atmospheric Programs. 2006. "Individual Emissions— Personal Emissions Calculator | Climate Change—Greenhouse Gas Emissions | U.S. EPA." October 19. https://www.epa.gov/ghgemissions/household-carbon-footprint -calculator.

Van Ryzin, G. G., S. Immerwahr, and S. Altman. 2008. "Measuring Street Cleanliness: A Comparison of New York City's Scorecard and Results from a Citizen Survey." *Public Administration Review* 68 (2): 295–303.

Veiga, Marcelo Motta. 2013. "Analyzing Reverse Logistics in the Brazilian National Waste Management Policy (PNRS)." In *Sustainable Development and Planning VI*, ed. Carlos Alberto Brebbia, 173:649–659. Ashurst, UK: Wit Press.

Verein f. mehr Demokratie. 2012. "Bundesweites Vorbild: Hamburger Transparenzgesetz Tritt in Kraft," October 5. http://web.archive.org/web/20131117202806/ http://www.mehr-demokratie.de/6033.html.

Verplanke, Jeroen, Javier Martinez, Gianluca Miscione, Yola Georgiadou, David Coleman, and Abdishakur Hassan. 2010. "Citizen Surveillance of the State: A Mirror for eGovernment?" In *What Kind of Information Society? Governance, Virtuality, Surveillance, Sustainability, Resilience*, ed. Jacques J. Berleur, Magda David Hercheui, and Lorenz Hilty, 185–201. Berlin, Heidelberg: Springer.

Vijayan, K. C. 2008. "Fugitive Ex-Citiraya Boss Fled with US$51m," August 17. http://news.asiaone.com/News/the%2BStraits%2BTimes/Story/A1Story20080815 -82432.html.

Voorhees, Erik. 2012. "@yanisvaroufakis @HostFat It Is Regulated, Only by Mathematics instead of Politicians.," October 1. https://twitter.com/ErikVoorhees/ status/252924410992943104.

Vos, P., E. Meelis, and W. J. Ter Keurs. 2000. "A Framework for the Design of Ecological Monitoring Programs as a Tool for Environmental and Nature Management." *Environmental Monitoring and Assessment* 61 (3): 317–344.

Voss, Jakob. 2007. "Tagging, Folksonomy & Co—Renaissance of Manual Indexing?" *arXiv:cs/0701072*, January. http://arxiv.org/abs/cs/0701072.

Wäger, P. A., M. Eugster, L. M. Hilty, and C. Som. 2005. "Smart Labels in Municipal Solid Waste—a Case for the Precautionary Principle?" *Environmental Impact Assessment Review* 25 (5): 567–586.

Warde, Beatrice. 1955. "The Crystal Goblet or Printing Should Be Invisible." In *The Crystal Goblet: Sixteen Essays on Typography*, 11–17. London: Sylvan Press.

Warneke, B., M. Last, B. Liebowitz, and K. S. J. Pister. 2001. "Smart Dust: Communicating with a Cubic-Millimeter Computer." *Computer* 34 (1): 44–51. doi:10.1109/2.895117.

Washington State Department of Ecology. 2010. "Solid Waste and Recycling Data." http://www.ecy.wa.gov/programs/swfa/solidwastedata/.

Waste Management Northwest. 2015. "Columbia Ridge Recycling and Landfill." http://wmnorthwest.com/landfill/columbiaridge.htm.

Weberman, A. J. 1980. *My Life in Garbology*. New York: Stonehill Press.

Weiser, Mark. 1991. "The Computer for the 21st Century." *Scientific American* (September): 94–104.

Welch, R. 1980. "Monitoring Urban Population and Energy Utilization Patterns from Satellite Data." *Remote Sensing of Environment* 9 (1): 1–9.

Wheeler, John Archibald. 1990. "Information, Physics, Quantum: The Search for Links." In *Complexity, Entropy and the Physics of Information*, ed. Wojciech H. Zurek, 310–336. Redwood City, CA: Westview Press.

WIEGO. 2013. "Waste Pickers—the Right to Be Recognized as Workers." Women in Informal Employment: Globalizing and Organizing. http://www.wiego.org/resources/waste-pickers-right-be-recognized-workers.

Wilson, David C. 2007. "Development Drivers for Waste Management." *Waste Management & Research* 25 (3): 198–207. doi:10.1177/0734242X07079149.

Wilson, David C., Ljiljana Rodic, Anne Scheinberg, Costas A. Velis, and Graham Alabaster. 2012. "Comparative Analysis of Solid Waste Management in 20 Cities." *Waste Management & Research* 30 (3): 237–254. doi:10.1177/0734242X12437569.

Winner, Langdon. 1980. "Do Artifacts Have Politics?" *Daedalus* 109 (1): 121–136.

Winner, Langdon. 2014. "Technologies as Forms of Life." In *Ethics and Emerging Technologies*, ed. Ronald L. Sandler, 48–60. Basingstoke, UK: Palgrave Macmillan.

Wisneski, Craig, Hiroshi Ishii, Andrew Dahley, Matt Gorbet, Scott Brave, Brygg Ullmer, and Paul Yarin. 1998. "Ambient Displays: Turning Architectural Space into an Interface between People and Digital Information." In *Cooperative Buildings: Integrating Information, Organization, and Architecture*, ed. Norbert A. Streitz, Shin'ichi

Konomi, and Heinz-Jürgen Burkhardt, 22–32. Lecture Notes in Computer Science. Berlin, Heidelberg: Springer. doi:10.1007/3-540-69706-3_4.

Woolgar, Steven. 1991. "Configuring the User: The Case of Usability Trials." In *A Sociology of Monsters: Essays on Power, Technology and Domination*, ed. John Law, 57–99. London: Routledge.

World Bank. 2016. *World Development Report 2016: Digital Dividends*. Washington, DC: World Bank.

Wurman, Richard S. 2000. *Information Anxiety 2*. 2nd ed. Indianapolis, IN: Que.

Wylie, Sara Ann, Kirk Jalbert, Shannon Dosemagen, and Matt Ratto. 2014. "Institutions for Civic Technoscience: How Critical Making Is Transforming Environmental Research." *Information Society* 30 (2): 116–126. doi:10.1080/01972243.2014.875783.

Zinnbauer, Dieter. 2012. "'Ambient Accountability'—Fighting Corruption When and Where It Happens." SSRN Scholarly Paper ID 2168063. Rochester, NY: Social Science Research Network. https://papers.ssrn.com/abstract=2168063.

Zinnbauer, Dieter. 2015. "Crowdsourced Corruption Reporting: What Petrified Forests, Street Music, Bath Towels, and the Taxman Can Tell Us About the Prospects for Its Future." *Policy & Internet* 7 (1): 1–24. doi:10.1002/poi3.84.

Zittrain, Jonathan. 2008. *The Future of the Internet—and How to Stop It*. New Haven, CT: Yale University Press.

Zuckerman, Ethan. 2013. *Digital Cosmopolitans: Why We Think the Internet Connects Us, Why It Doesn't, and How to Rewire It*. 1st ed. New York: W. W. Norton & Company.

# Acknowledgments

This project would not have been possible without the support, encouragements, and constructive criticism from a large number of friends and colleagues.

This book originated from my doctoral research, and I am thankful to my mentors at the Department of Urban Studies and Planning at MIT, which offered me an extraordinary intellectual environment for this project. I want to especially thank my PhD advisor and committee members Carlo Ratti, Lawrence Vale, Eran Ben-Joseph, and Brent Ryan—I am fortunate and honored for benefiting from their exceptional knowledge and support. My gratitude also goes to faculty members at MIT whose generous intellectual support has shaped my work, including Frank Levy, Rex Britter, Steve Miles, Joseph Ferreira, Karen Polenske, and late JoAnn Carmin, who helped me at a pivotal moment of my journey.

I want to thank my editors Jay Slagle and Lucas Freeman for their help developing the manuscript and making it accessible; and the series editors Geoffrey Bowker and Paul Edwards, and Katie Helke at MIT Press for their support and patience in completing this book.

The Trash Track project would not have been possible without all the volunteers who participated in the project and generously gave us their time, thoughts, and enthusiasm. I am fortunate to have been able to work on an extraordinary team including David Lee, Malima Wolf, Avid Boustani, Jennifer Dunham, and Assaf Biderman who all made this project a delightful experience. I also want to thank my colleagues at the Senseable City Lab, Kristian Kloeckl, Tony Vanky, Clio Andris, and Kael Greco. My gratitude goes as well to Jim Puckett, who gave me an excellent first-hand insight into the world of waste forensics and environmental advocacy.

The Forager project was generously supported by the Carroll L. Wilson Foundation, MISTI Brazil, and MIT Global Challenge. I am thankful to my friends and colleagues who shaped the project with their ideas, contributions, and continuing support, first of all, David Lee and Laura Fostinone, who influenced the project from the very beginning, Lucia Helena Xavier, Julian Contreras, and Rafael Galvão. Special thanks go to the mentors of Forager, Maria Cecilia Loschiavo dos Santos and Libby McDonald, and all partners and cooperative members in Brazil.

With regard to my work with urban feedback systems, I would like to thank Chris Osgood and Nigel Jacob from the Boston Office of New Urban Mechanics for their support.

I am grateful to my friends Orkan Telhan, Duks Koschitz, Susanne Seitinger, Peter Schmidt, and Katja Schechtner. My family—Karl, Elisabeth, Martin, Ibrahim, Munira, Nadja, Enes—you are my foundation. Most importantly I thank my wife Azra Aksamija, who has always stood by my side with spirited ideas and loving kindness.

# Index

Printed in the United States
by Baker & Taylor Publisher Services